The Antarctic Subglacial Lake Vostok

Glaciology, Biology and Planetology

Igor A. Zotikov

The Antarctic Subglacial Lake Vostok

Glaciology, Biology and Planetology

Published in association with
Praxis Publishing
Chichester, UK

PRAXIS

Dr Igor A. Zotikov
Russian Academy of Sciences Fellow (correspondent member)
Institute of Geography of Russian Academy of Sciences
Moscow
Russia

SPRINGER–PRAXIS BOOKS IN GEOPHYSICAL SCIENCES
SUBJECT *ADVISORY EDITOR*: Dr. Philippe Blondel, C.Geol., F.G.S., Ph.D., M.Sc., Senior Scientist, Department of Physics, University of Bath, Bath, UK

ISBN 10: 3-540-42649-3 Springer-Verlag Berlin Heidelberg New York

Springer is part of Springer-Science + Business Media (springeronline.com)

Bibliographic information published by Die Deutsche Bibliothek

Die Deutsche Bibliothek lists this publication in the Deutsche Nationalbibliografie; detailed bibliographic data are available from the Internet at http://dnb.ddb.de

Library of Congress Control Number: 2006926430

Printed in Germany

Cover design: Jim Wilkie
Project management: Originator Publishing Services, Gt Yarmouth, Norfolk, UK

Printed on acid-free paper

Contents

Preface

It seems that nearly all my life has been associated one way or another with Vostok Station and the Lake beneath it, now known as Lake Vostok. Vostok Station was opened in 1957, 35 years before the discovery of Lake Vostok was officially declared. I came to Vostok Station in 1958 working on a project to drill into the ice sheet, or more accurately, to melt a borehole with the intent to penetrate deep enough to find the temperature difference at the bottom of the ice sheet. Unfortunately, to my dismay, my "drill" broke down at a depth of 50 meters. I thought then that with some luck, we will penetrate the ice sheet by simply melting through the ice column in 2–3 years. I would not have believed then, in 1958, that we would be drilling the ice sheet for more than 30 years and still not reach the bottom.

Following my graduation from the Moscow Aviation Institute in the Soviet Union in 1949 I worked for three years on the heat transfer and thermodynamic problems of designing jet engines. Following that I worked, from 1952 until 1958, on the re-entry problems of the first Soviet ballistic missile, mainly related to melting/evaporation of its nose by the heat that was generated by its return to Earth. My Ph.D. in this field in 1957 coincided with the beginning of the International Geophysical Year (IGY) and Soviet Union participation in the study of Antarctica. The possibility to go to Antarctica and see the largest glacier on Earth and also view the world from outside the Soviet Union resulted in my application to work there as a "thermo-physics researcher" in the glaciological science branch of the expedition. Upon my application being accepted, I learned about Vostok Station and the prospect of working on the thermal regime of the Antarctic Ice Sheet.

In 1960–1963 I published papers that showed the temperature at the bottom of the ice sheet below Vostok Station to be at the ice melting point, as well as being beneath the thickest part of the ice sheet. It occurred to me that lakes might exist at the ice/rock interface, and that some kind of life should exist there. There did not seem to be much interest in the prospect of life then, in 1963, although my publications were taken seriously and translated and published in English. As a result, I suddenly and unexpectedly became a glaciologist, and I determined that my future would be devoted to Antarctica.

In 1963 I re-visited Vostok Station as a member of Dr. Kapitsa's team to participate in his traverse, during which seismic soundings of the ice sheet showed reflections from the top and bottom surfaces of what is now known as Lake Vostok. However, none of us would hear of confirmation of its existence for a further 30 years.

In 1965 I spent my second overwinter in Antarctica at the main American research station, McMurdo, as a Soviet exchange scientist and member of the U.S. "Deep Freeze 65" Antarctic Expedition. I studied the processes of freezing and melting at the bottom of the Ross Ice Shelf and met leading American specialists on deep drilling of the ice sheet. I summarized the results of my Antarctic expeditions in a DSc. dissertation on glaciology in 1967, with an emphasis on bottom melting of a large area of the central part of the Antarctic Ice Sheet with the presence of meltwater at the bottom. However, few glaciologists agreed with this concept at this time, and for that reason I felt that I could not defend my dissertation successfully. Fortunately, my American colleagues at that time were drilling to the bottom of the ice sheet at Byrd Station (80°S, 120°W) and found liquid water at the bottom. My driller friends informed me of their find by telegram in the Soviet Union, and I then presented the news to the Science Council of the Arctic and Antarctic Research Institute in Leningrad.

In 1972 I was invited by the U.S. National Science Foundation (NSF) to join the Ross Ice Shelf Project (RISP), becoming a Principal Investigator to determine whether freezing or melting occurs at the bottom of the Ross Ice Shelf, working on this project until 1978 using McMurdo as a base. McMurdo was also the base for Dr. G. de Q. Robin's science team using a U.S. Navy LC-130 aircraft for radio-echo sounding to locate subglacial lakes below the East Antarctic Ice Sheet. Some evidence was found during those surveys of a large subglacial lake near Vostok Station.

Dr. Robin and I knew each other as a result of a common interest in the existence of subglacial lakes and, as a result, I participated in flights to find surface evidence of the lake beneath Vostok Station. In 1993, after we confirmed evidence of Lake Vostok's existence, I was in Cambridge, U.K., at the Scott Polar Research Institute assisting Dr. Robin to organize the first, and later, second workshops on the study of Lake Vostok, resulting in a coauthored article published in *Nature*. The lake thus became publicized for future studies by a wide scientific community.

Lake Vostok is a remarkable feature in that it serves a multidisciplinary group of studies in glaciological, geophysical, hydrological, biological, and planetary research. Located under 4,000 m of ice in the interior of East Antarctica, it occupies a depression in the bedrock beneath Vostok Station. The lake is about 250 km long and 50 km wide, comparable with Lake Ontario in North America. Seismic studies suggest that the lake is as much as 1,000 m deep. It is considered by many that the discovery of Lake Vostok is among the most important geographical discoveries of the second half of the twentieth century.

The bottom of the ice is at the pressure melting point, warmed by geothermal heat. The amount of trapped heat is small, but the lake is too far below the ice sheet surface for the penetration of low temperature extremes of the Antarctic atmosphere. This part of the Antarctic Ice Sheet has remained relatively stable for many hundreds of thousands of years, perhaps in excess of a million years, although it has thickened and thinned in response to global and regional climatic changes.

An intensive drilling campaign has been carried out for about 30 years by Russian scientists at Vostok Station for ice cores for climate research. As the drilling achieved greater depths the possibility of reaching the ice sheet base became a reality. However, concerns were raised by the scientific community about the consequences of drilling through the ice and entering the lake. For example, what would happen when the drill reaches the water? Does the lake contain uncontaminated water and micro-organisms? What sort of biological life might exist at 4,000 m beneath the ice, isolated from other influences for perhaps a million years? These issues led to several international workshops on Lake Vostok to consider the wider implications, a possible science plan for the drilling and sampling of the lake water, and other necessary investigations and technical developments.

The aim of this book is to collate material about Lake Vostok and to organize and synthesize it to establish its complete geographical picture. I hope that this will help to establish a new understanding of the different and still hidden aspects of the phenomenon for the interest of the general public and scientists.

The following topics are covered in the book:

(1) Glaciology of the central part of the East Antarctic Ice Sheet is discussed, including the related aspects of the heat balance at the bottom of the ice sheet; critical thickness, bottom melting and subglacial lake concepts; the existence of life in subglacial water (glaciological reasons); and the average bottom water layer thickness.
(2) The problem of Lake Vostok as one of many other subglacial lakes is examined. It includes a summary of radio-echo sounding surveys of subglacial lakes; seismic determinations and their interpretation; satellite altimetry data and other surface evidence of subglacial lakes; a history of the discovery and the type of ice cover of Lake Vostok (internal Lake Vostok Ice Shelf concept); and the salinity and currents in the water of the lake.
(3) Modern data on biological inclusions in the Vostok ice core, and the ecology of Lake Vostok water and its living creatures (micro-organisms) are critically reviewed. The scientific importance of life in such a lake is discussed, along with the possibility that Lake Vostok organisms, if any, might present a danger for the rest of the world.
(4) Applications to planetary studies include plans to penetrate the ice cover of Europa (a moon of Jupiter) for subglacial water, as well as subglacial lakes beneath Martian ice.
(5) Penetration into the lake is discussed as a major topical problem. Topics include a comparative analysis of methods of penetration through the ice sheet; a history of deep drilling at Vostok Station; problems of the possibility of contamination

of organisms on penetration into Lake Vostok; and autonomous nuclear powered subglacial stations and the NASA "Cryobot" concepts for use in Lake Vostok and for planetary studies.

At the time of writing this book (January 2006) Lake Vostok had not been drilled into. The most exciting and important part of this scientific story will come to us in 3–5 years when we, hopefully, obtain samples of the water from the lake, investigate the water thickness via remote systems, and discover any implications. Experimental data on the salinity and currents in the lake's water, and data on biological material, if present, will possibly change the interpretation of their existence. These events might happen relatively soon (2010–2015), but this book will serve as a prelude to what will surely result from studies of this remarkable feature.

It is uncertain whether British, American, Russian or scientists from other countries will be first to penetrate the ice sheet and probe the lake, but approaches to the penetration in Russia are currently different, depending on the country.

During recent years I worked intensively on the issues of Lake Vostok in Russia, with support from the Russian Foundation for Basical Research (RFBR). I received awards for a 3-year grant in 1996 for "Lake Vostok Studies", a 2-year grant in 1999 for "An alternative approach to Lake Vostok Genesis", and a 2-year grant in 2001 for further studies. Work on these grants generated a large amount of material in Russian, and I became a member of a committee for the evaluation of Russian drilling equipment for the purpose of penetrating the lake. I thank the RFBR for their support of this work.

The idea to write a book about Lake Vostok was proposed to me by Clive Horwood of PRAXIS Publishing Ltd (U.K.) in 1998 following a suggestion by one of his leading authors, Academician Kirill Kondratyev. It was planned to be a strictly scientific book in the beginning, but in the process of writing I yielded to the idea of a content that would be more popular in nature, essentially a "drama of ideas". This change, however, delayed the completion of the draft manuscript.

In the autumn of 2003 I received a grant from the U.S.A. Fulbright Foundation for 8 months of work in the U.S.A. on a project called "Geographical study of Lake Vostok (Antarctica) based on glaciological, geophysical, hydrological, biological, and planetary data analyses". I left the Institute of Geography of the Russian Academy of Sciences, for the U.S.A. to conduct the project under the auspices and direction of Dr. Roger Barry, my friend and colleague for more than 25 years, and Director of the National Snow and Ice Data Center (NSIDC) of the Cooperative Institute for Research in Environmental Sciences of the University of Colorado at Boulder. My work there began in January 2004, and with Dr. Barry's enthusiastic support, my first draft was completed in September 2004.

Figures

Abbreviations and acronyms

AARI	Arctic and Antarctic Research Institute
AMANDA	Antarctic Muon and Nutrino Detector Array
ANARE	Australian National Antarctic Expedition
BAS	British Antarctic Survey
CEE	Comprehensive Environmental Evaluation
ECM	electrical conductivity measurement
EIA	Environmental Impact Assessment
EPICA	European Project for Ice Coring in Antarctica
ERS	European research satellite
ESF	European Science Foundation
GPHS	general purpose heat source
GPS	global positioning system
IGY	International Geophysical Year
IPY	International Polar Year
ISAIRAS	International Symposium on Artificial Intelligence, Robotics and Automation in Space
JPL	Jet Propulsion Laboratory
KEMS	Russian electro-mechanical drill for coring (Russian abbreviation)
LBR	low bit rate
LVHRZ	Lake Vostok Hypothetical Rift Zone
MOLA	Mars Orbiter Laser Altimeter
NASA	National Aeronautics and Space Administration
NSF	National Science Foundation
NSIDC	National Snow and Ice Data Center
PCR	polymerase chain reaction
PLAS	Sub-glacial Autonomous Station (SGAS) (Russian abbreviation)
PMGRE	Polar Marine Geological Research Expedition (Russian)
RAE	Russian Antarctic Expedition

RES	radio-echo sounding
RFBR	Russian Foundation for Basical Research
RIA	Russian Information Agency
RISP	Ross Ice Shelf Project
RMS	root mean square
RSS	reflection seismic sounding
SAE	Soviet Antarctic Expedition
SALEGOS	Special Antarctic Lake Exploration Group of Specialists
SAR	synthetic aperture radar
SCAR	Scientific Committee on Antarctic Research
SEM	scanning electron microscope
SGARS	Sub-glacial Autonomous Returnable Station
SGAS	Sub-glacial Autonomous Station
SPRI	Scott Polar Research Institute
STAIF	Space Technology and Application International Forum
TBZS	Russian thermo-electric drill for holes filled with liquids (Russian abbreviation)
TELGA	Russian thermo-electric drill for dry holes (Russian abbreviation)
TUD	Technical University of Denmark
USARP	U.S. Antarctic Research Program
WDC	World Data Center

Acknowledgements

I am grateful to numerous individuals who assisted me in this study. To start I thank Academician G. A. Avsiuk, who hired me at the Institute of Geography of the Soviet (later Russian) Academy of Science, the Institute Director Academician V. M. Kotlyakov and Institute personnel I worked with since 1962 (especially Dr. V. Barbash, Dr. N. Osokin, Dr. J. Raikovsky, and Dr. V. Zagorodnov); the Director of the Arctic and Antarctic Research Institute in St. Petersburg, Academician A. F. Treshnikov, all those at the Institute and members of Soviet and Russian Antarctic Expeditions I worked with (especially Dr. N. Barkov, Prof. A. Kapitsa, Dr. V. Lipenkov, Dr. V. Lukin, Dr. L. Savatugin, and Mr. V. Morev). I thank all members of the U.S. Antarctic Research Program beginning with "Operation Deep Freeze 65" and the overwinter party of 1965 at McMurdo Station (especially Cdr. D. Blades, Mr. H. Ueda, and Dr. A. Gow); those at the Office of Polar Programs of the National Science Foundation in Washington, D.C. (especially Dr. R. Cameron, Dr J. Fletcher, and Cdr. R. Dale); the Ross Ice Shelf Office at the University of Nebraska–Lincoln (Prof. R. Rutford, Mr. B. Lyle Hansen, and Mr. K. Kuivinen); and field parties at McMurdo Station and J-9 camp from 1972 to 1978 (especially Mr. S. Jacobs). I would like to thank those at the U.S. Army Cold Regions Research and Engineering Laboratory at Hanover, New Hampshire, for their hospitality, where I stayed for 6 months in 1979 as a visiting scientist; the people of the Ice Core Facilities, State University of New York, where I worked with Dr. C. C. Langway for 6 months in 1983; those at Lamont–Doherty Geological Laboratory (now Earth Observatory) (especially Mr. S. Jacobs), where I stayed for 3 months in 1986; and Professor O. Watanabe, who invited me to work for 6 months on this project at the Japanese Antarctic Research Institute in Tokyo, Japan, in 1995 – as well as everybody at this Institute.

My special thanks go to Drs. G. de Q. Robin, J. Dowdeswell, M. J. Siegert and their colleagues at the Scott Polar Research Institute of Cambridge University, U.K.

Dr. Robin invited me in 1993 and 1996 to work on the problems of Lake Vostok, and this collaboration became one of the most important contributions to the book.

My very special thanks go to Dr. R. Barry, Professor of Geography and Director of the National Snow and Ice Data Center (NSIDC) and World Data Center (WDC) for Glaciology of the Cooperative Institute for Research in Environmental Sciences of the University of Colorado in Boulder and host for my 8 months Fulbright Foundation Scholarship in 2004 in the U.S.A. Without his generous suggestion that I should use the stay for writing this book, and without his supervision, this task would not have been completed.

I also want to include everyone at the NSIDC, WDC, and other institutions in the U.S.A. for the permanent support and friendship I received during my 8-month stay. Finally, I thank Dr. R. H. Rutford, Excellence in Education Professor of Geosciences at the University of Texas in Dallas. He was actively involved in Antarctic studies for the whole period described in the book, was President of the Scientific Committee on Antarctic Research (SCAR) for many years dealing with issues related with the study of Lake Vostok, and supported the idea for this book from the beginning, being very generous with his time for discussions. Professor Rutford helped me with this book in one further way. He asked his friend, a known American editor, polar explorer, and the President of the American Polar Society (2005), John Splettstoesser, to assist me editorially in a critical review of the manuscript converting the text into a form suitable for publication – John agreed to do this. His generosity, willingness, and personal desire to see this book in print made it possible. Dr. Philippe Blondel (University of Bath), scientific editor for Praxis Publishing Ltd edited it comprehensively too. Thank you John and Philippe!

1

From Lake Vostok to Europa: Tractors and satellites

A fool always wants to shorten space and time; a wise man wants to lengthen both

Words of J. Ruskin in *Modern Painters*, which Dr. E. A. Wilson, a member of Captain R. F. Scott's Antarctic expedition, wrote in his diary on "Thursday 11 December 1902" on their way to the South Pole (Wilson, 1982)

It is a long story, more than 40 years old, of a scientific (and a human) struggle where "Murphy's Law" and just plain good luck were competing against one another, with the sequence of events however moving in the right direction.

Let us start with the journal *Nature*, 20 June 1996, which illustrated a large map of Antarctica on the cover. A feature in red appeared at the center of the map, and below the map, in big letters, were the words "Giant lake beneath the Antarctic Ice" (Figure 1.1).

There was a description of the picture within the magazine on the contents page, stating that some traces of the giant lake were discovered below the East Antarctic Ice Sheet 20 years ago, and now new laser altimetry and radio-echo sounding data proved that the largest of all known subglacial lakes was discovered. The size of the lake approached that of one of the Great Lakes of North America, and was only a few times smaller than Lake Baikal. Its length was estimated at about 200 km, its width about 50 km, and its depth about 500 m. Picturesque details on the front cover figure of *Nature* (Bamber, 1994) showed the Antarctic Ice Sheet surface, developed at the Millard Space Science Laboratory of the University College of London on the basis of altimetry data obtained from ERS-1, the first European satellite designed to study the Earth (Ridley *et al.*, 1993; Bamber, 1994).

The data, when combined with known ice sheet elevations, produced surface features accurate to within a few meters, showing the lake's presence at several thousand meters below the surface. The lake was named Vostok, after the name of the Russian Antarctic research station located above it.

About 40 years before the article in *Nature* was published, the International Geophysical Year (IGY, 1957–1958) was in its planning stages, and major

nature

INTERNATIONAL WEEKLY JOURNAL OF SCIENCE

Volume 381 No. 6584 20 June 1996 £4.00 FFr44 DM17.5 Lire13000 A$12

Giant lake beneath the Antarctic ice

Figure 1.1. Lake Vostok on the front page of *Nature*, June 1996.

developed countries of the time were involved in finding locations for their research stations in Antarctica. Most countries agreed on coastal sites in order to reduce logistical problems, but both the U.S.A. and the Soviet Union also planned to position stations in the middle of the Antarctic Ice Sheet.

At the start of these preparations, the Soviet Union had shown no interest in the continent since Bellingshausen's circumnavigation in the early nineteenth century. As delegates of the eleven nations assembled in Paris in July 1955 to plan their cooperative effort, it was not known whether a Soviet representative would join them. "I know that he [the Soviet] was waiting for permission to go abroad, which was difficult to get at this time. He got it too late and arrived with delay." (Quigg, 1983). It became clear that the Soviet Union planned a major program – at least three bases (soon to be six), including one at the South Pole. Since the U.S.A. had already expressed a prior interest in the site, there was considerable relief when the Soviet delegate deferred without argument and chose instead the Pole of Relative

Inaccessibility (the spot in the middle of Antarctica, which is located at the longest distance from any shore) and the Geomagnetic Pole (Quigg, 1983).

As a result, a close connection of Russian scientists to Lake Vostok was established because of this delay in July 1955. The history behind the events at Lake Vostok is worth relating – one reader of the first draft of this manuscript wrote "Being published in one of the high-ranking scientific journals does not mean that this picture (and previous work) was the beginning, the pinnacle, or the end of work on Lake Vostok. In my opinion, it is an excellent idea to remind people, scientists or not, that events did not start when interest was shown by NASA in relatively recent years, but had in fact been going for half a century."

According to the agreement in Paris the Soviet Antarctic Expedition (SAE) constructed its main scientific station in 1956 at Davis Coast in East Antarctica, the closest coastal location (about 1,300 km) from the Geomagnetic Pole in the interior. The station was named Mirny, which means "peaceful", after the name of one of the two ships of the first Russian Antarctic Expedition of 1820. In the next year, ten large Kharkovchanka tractors left Mirny Station, towing insulated huts on sledges and more than 100 tons of diesel oil. About 90 days and nights later, after a non-stop struggle, 30 tons of diesel fuel and 3 tons of food provisions were brought to the interior site, enough to establish an overwinter station for four people.

The huts, on sledges, were moved close together, forming a compact four room house. A generator station, kitchen, and a meteorological laboratory were erected, as well as radio masts and a pole for the station flag. An airstrip on snow, capable of accommodating twin-engine Russian Ilyuchin-2 aircraft in an Antarctic summer, was constructed and a new year-round station began operation.

This station was named Vostok, which means "east" in Russian. It was also named after the second ship of the First Russian Antarctic Expedition of 1820. Located at 3,500 m above sea level, it became the coldest place on the Earth's surface to be occupied by human beings – the lowest temperature registered was −89°C (*Atlas of Antarctica*, 1966).

No one knew at this time that there were about 4 kilometers of glacier ice below the station, and that this ice sat above the surface of a large subglacial lake, with water some half a kilometer deep. There was only one such lake of this size in the ten million square kilometers of the surface of Antarctica, and only two interior stations. Of these two stations, one just happened to be located precisely above the lake. The likelihood of such an occurrence is remarkable. Figure 1.2 shows the position of Vostok Station.

Kapitsa *et al.* (1996) published what the *Nature* issue called the discovery of a large, deep freshwater lake beneath the ice sheet, while others (Ellis-Evans and Wynn-Williams, 1996; Bentley, 1996), explained a mechanism for the water formation and reasons for its existence. The mechanism was simple – the water formed as a result of the entrapment of the geothermal heat flow. The possibility of some forms of life in the lake was also discussed. *The Times* newspaper devoted a special article to this event on the same day that the *Nature* issue was printed (Nuttal, 1996). The author of this article reviewed the new details of the lake with a freshwater source deep below the ice, commenting that it reminded him of a famous Jules

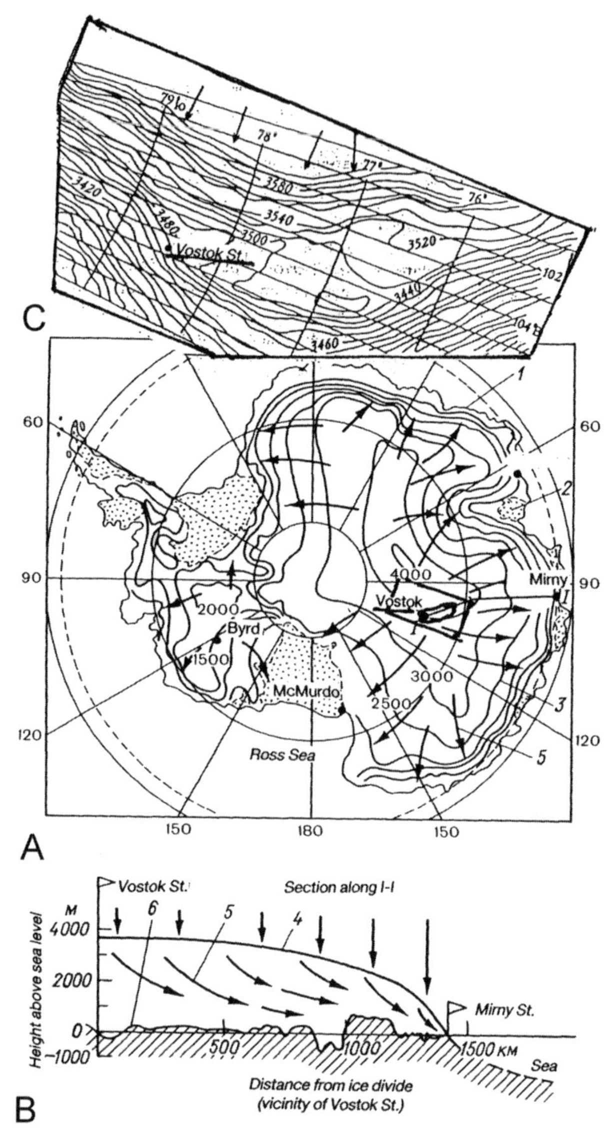

Figure 1.2. Schematic map of the East Antarctic Ice Sheet, plan view (A), showing Lake Vostok and Vostok and Mirny Stations and a cross section along I–I (B). In (A) note the following: (1) outer contours of the ice sheet; (2) ice shelves (floating parts of the ice sheet); (3) isolines of the elevation of the ice sheet surface above sea level; (4) upper surface of the ice sheet along the line I–I; (5) directions of ice movement; and (6) ice sheet–bedrock interface along line I–I (adapted from Zotikov, 1986). An enlargement of the map in the vicinity of Vostok Station is shown in (C). Surface elevation contours for the enlargement are from Ridley *et al.* (1993). A flat nearly horizontal surface of ice marks Lake Vostok surrounded by the relatively steep slopes of the ice sheet. Arrows show the direction of ice flow from a nearby ice divide.

Figure 1.3. A large, hidden subglacial lake is mindful of Jules Verne's lake in *Journey to the Centre of the Earth*, where underground travelers find a lake containing sea monsters. *The Times* published this picture from Verne's novel the same day that the *Nature* issue about Lake Vostok was published (Nuttall, 1996).

Verne science fiction novel, in which individuals discovered a huge lake deep underground, and in the water of this lake they found sea monsters, which existed only there, remaining unknown to science (Figure 1.3). Contrary to Verne's lake, Lake Vostok looks like a cup, with the top cap being a thick ice cover. This cover keeps the lake separated from the rest of the world.

Biologists suggested that Lake Vostok might be a location for living microbes, and because they have had no contact with life in the rest of the world for a lengthy time period, they would be different from anything that lives now. It is possible that they could be used as genetic material for new developments in the medical treatment of humans and animals, as well as for other industrial applications.

Other newspapers and magazines in the U.K. also published articles about Lake Vostok on 20 June 1966, repeating the concept of "The Lost World" (Ratford, 1996) and "The Time Machine Capsule". *New Scientist* magazine projected into the future, and mentioned that recovery of uncontaminated live micro-organisms from the lake was comparable with that of bringing biological samples from Mars to the Earth (Muir, 1996).

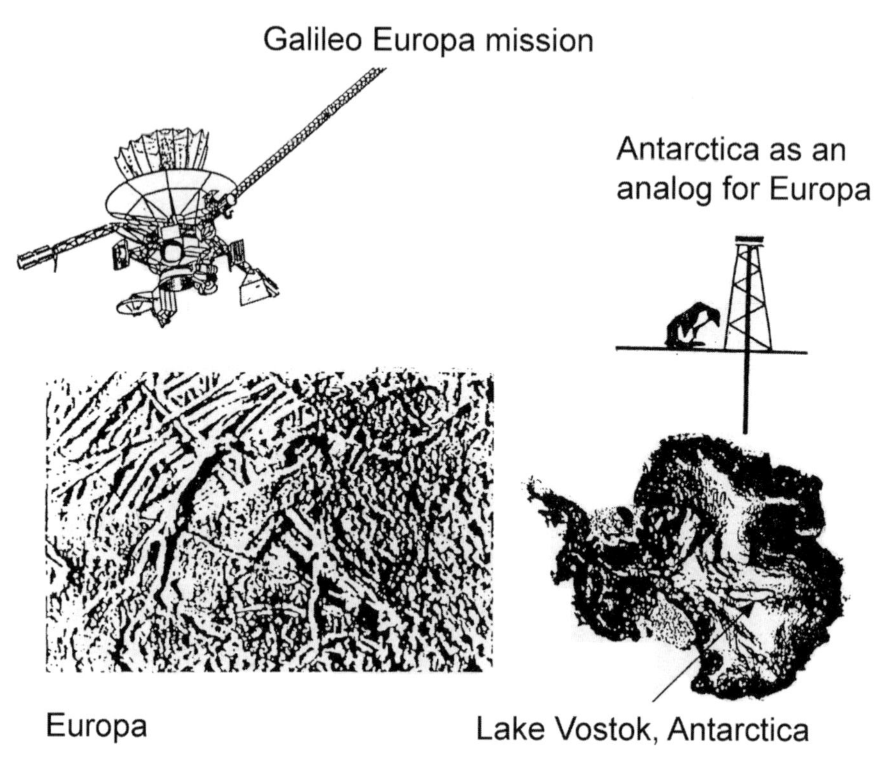

Figure 1.4. Subglacial Lake Vostok as an analog of a sub-ice ocean on the Jovian moon Europa (Carsey and Horvath, 1996).

Publicity about the discovery of a giant lake, separated from the rest of the world by many kilometers of ice "armor", raised excitement and interest for many scientists and engineers involved in the study of Solar System planets, especially Europa, one of the Jovian moons.

The publication in *Nature* in 1996 on the discovery of Lake Vostok appeared at the time when information from the Galileo spaceship was arriving back on Earth (workshop at the end of 1996), proving that the surface of Europa (with its diameter of more than 3,000 km) is covered with a sheet of ice (Carsey and Horvath, 1996) (Figure 1.4).

The thickness of this ice sheet is estimated at this time to be several kilometers to hundreds of kilometers, and it appears to be a system of large subglacial lakes or a sea of liquid water below ice, similar to Lake Vostok on Earth. The existence of liquid water below Europa's ice sheet is still not certain, but has been inferred from measurements of gravity, moments of inertia, and study of images of the moon's surface. But the probable cause of sub-ice water is the same as for Lake Vostok – heat flow from beneath the ice sheet. However, the source of heat in the case of Europa is different. Gravity effects due to Jupiter itself and its moons Io and Ganymede produce considerable heat deep within the interior of the moon, and the thermal insolation of a thick ice layer is substantial. Liquid water below the ice sheet should exist, in spite of the temperature at the surface of the moon, estimated at about −170°C. A logical assumption is that if there is life in Lake Vostok, perhaps there is also life in the sub-ice ocean of Europa, at least in its simplest forms.

2

Water below the central part of the ice sheet

Many years before the literature discussed in Chapter 1 was published there was related information on locations where liquid water should exist beneath ice sheets.

It was believed, at the end of the 19th century, that the temperature at the bottom of an ice sheet increased with increasing ice thickness. P. A. Kropotkin, a Russian Duke and co-founder of the world's anarchy movement was also a scientist. In a book on the Quaternary glaciation (Kropotkin, 1876), he postulated that the heat regime within thick, cold ice sheets, below the layer of ice exhibiting seasonal temperature changes, was at a steady state, with temperature increasing linearly with depth, allowing the geothermal heat flow to move toward the surface of the ice (assuming that the vertical movement within an ice sheet is negligible).

N. N. Zubov, a Russian "oceanologist", a rear admiral in the Soviet Navy, Professor at the Moscow State University, and a leading world specialist on Arctic sea ice, used this approach for analyses of thermal conditions at the bottom of cold, thick Antarctic-type glaciers (Zubov, 1955, 1956, 1959). If the temperature increases linearly with depth and the gradient is governed by a need to pass all the geothermal heat flow through the ice cover by thermal conductivity only, then there is a critical ice thickness, corresponding to the bottom temperature of the ice sheet, equal to an ice melting point (Figure 2.1(A)). Figure 2.1(B) corresponds to a thin glacier, in which the bottom temperature is below the freezing point of water. This case corresponds to the existence of permafrost below the ice.

This approach led Zubov to conclude that the ice sheet thickness cannot be greater than its critical thickness. A greater thickness in Zubov's model would correspond to the bottom temperature being higher than the melting point of ice. However, he was the first to publish the fact that the thickness of the Antarctic Ice Sheet in some locations, as determined from seismic soundings, was much greater than his critical thickness (Zubov, 1956). As a result, he introduced a bottom layer, with a thickness equal to the difference between the measured thickness and the calculated critical one (Figure 2.1(C)). He attributed this layer, with some surprise

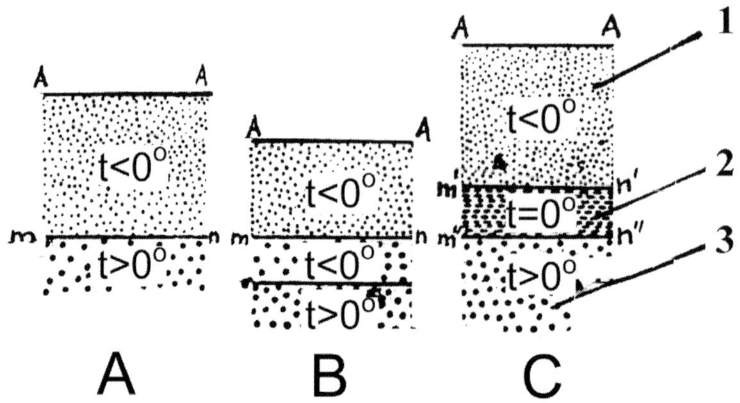

Figure 2.1. A–A represents the upper surface of the ice; 1 – ice thickness; 2 – "some mixture of water and ice"; 3 – bedrock, as proposed by Zubov. (A) The thickness of the ice is equal to "the critical thickness" corresponding to an ice bottom temperature equal to the ice melting temperature. (B) "Thin" ice, in which the bottom temperature is less than the ice freezing point. This case represents the existence of a layer of permafrost below the ice. (C) "Thick" ice, with a thickness larger than the critical thickness. In Zubov's model (C), some bottom layer should exist, in which the thickness is equal to the difference between a measured thickness and a calculated critical one. Zubov thought that it "could be some mixture of water and ice..." (Zubov, 1956, p. 26).

to himself, to a mixture of water and ice (Zubov, 1956). Figure 2.2 shows Professor Zubov at about the time he wrote this paper about the critical thickness of Antarctic ice.

In 1959, A. P. Kapitsa, a young Russian geographer also from Moscow State University, used Zubov's approach to suggest the existence of liquid water lenses below the ice in places identifying with Figure 2.1(C) in central parts of the East Antarctic Ice Sheet (Kapitsa, 1961). However, soon after this, a new approach to an understanding of the temperature distribution in cold, Antarctic-type ice sheets appeared – one that showed more accuracy than Zubov's model.

In 1955, Dr. Gordon de Q. Robin, a new name in glaciology at that time, a former first mate on a World War II submarine in the Australian Navy, and a graduate physicist of Melbourne University, published his first and famous article giving a theoretical explanation of the vertical temperature distribution through the Maudheim Ice Shelf, Antarctica, and the Greenland ice cap near Camp Century (Robin, 1955). This article resulted from his participation as a member of the Norwegian–British–Swedish Antarctic Expedition (1949–1952) based at Maudheim Station on the Maudheim Ice Shelf, a floating mass of ice some 200 m thick. He found that the ice temperature increased from a mean annual temperature near the surface to that of the freezing temperature of seawater at the bottom of the ice shelf. However, this change was not linear, as Zubov suggested. The temperature increased very slowly with depth in the upper part of the ice shelf, and then changed rapidly in the lower part as it approached the bottom of the ice shelf. Dr. Robin showed that a

Figure 2.2. Professor Zubov in 1955, when he wrote his papers on the critical thickness of Antarctic ice.

measured profile corresponded to a theoretical profile in a steady-state system, with temperature, surface accumulation, and ice thickness assumed to be constant. This temperature profile is valid if a thermal conductivity equation includes both thermal conductivity and vertical heat convection terms, due to the vertical movement of ice from the surface to the bottom of the ice due to continuous accumulation at the surface and melting of the ice shelf at the bottom.

This approach accurately explained an experimental temperature distribution for a 1,200 m thick ice sheet in Greenland near Camp Century. The article written by Dr. G. de Q. Robin remained for many years a classic work in glaciology. Figure 2.3 shows Dr. G. de Q. Robin at the time he wrote this paper.

This approach, as in those used by Kropotkin and Zubov, also used a steady-state approach to calculate temperature conditions at depth in ice sheets of Antarctica and Greenland. However, it also required that two terms of the energy equation be included, thermal conductivity and the vertical component of convection, resulting in a non-linear vertical temperature distribution in the ice sheet, making it possible to achieve a good correlation between calculated (measured) profiles and experimental data.

This meant that heat transfers vertically not only by heat conductivity, but by continuous vertical movement of cold particles of ice from the surface to the bottom of a glacier as a result of the deformation and spreading of ice particles due to

Figure 2.3. Dr. G. de Q. Robin at the XIII Scientific Committee on Antarctic Research (SCAR) meeting at Jackson Hole, Wyoming, U.S.A. in 1974, at the time of his radio–echo sounding discovery of subglacial lakes.

gravity, and a continuous replacement of ice by accumulation of snow at the surface, leading to a vertical convection of cold from the surface to internal, warmer parts of the ice sheet. Dr. Robin correctly explained the experimental temperature distribution for the 200 m thick Maudheim Ice Shelf and also for the 1,200 m thick ice sheet in Greenland near Camp Century. He suggested that the temperature distribution in these ice masses was formed under the influence of a geothermal heat flow from beneath the grounded ice sheet in Greenland, or because the bottom temperature was equal to the ice melting point in the case of a floating ice shelf.

All this heat flow for a steady-state approximation is transferred upward in the ice. Dr. Robin showed that the temperature at the bottom of the Greenland Ice Sheet near Camp Century, which was calculated using this method, was much higher than the constant temperature at the surface, but lower than the freezing temperature of the ice, illustrating a good correlation with measured temperature profiles.

It is interesting that if Robin applied his methodology to conditions in the central part of the Antarctic Ice Sheet, the calculated bottom temperature would be higher than the freezing temperature of water. Because this is physically impossible, a change in boundary conditions would be required for the ice–bedrock interface. He would then presumably include a term for a heat sink due to the permanent melting of ice at this boundary, and would make the melting rate sufficiently high so that the bottom temperature would be equal to the freezing point of water. However, Dr. Robin did not consider the case for a thick, grounded ice sheet, so did not see the need to include a term for bottom melting. Because of that omission, the phenomenon of permanent melting at the bottom of very thick ice sheets, and the possibility of subglacial lakes due to water accumulation was not apparent at that time.

When Dr. Robin published his important work on this subject, I knew very little about the Antarctic Ice Sheet. As a graduate of the Moscow Aviation Institute, I worked on Ph.D. experiments that were involved in research connected with solving missile "re-entry problems". Using a variety of materials with low melting points, I manufactured models of nose cones and cylinders and installed them in the hot jet streams of huge rocket engines or in hot supersonic airstreams using specially designed apparatus. I studied how these supersonic shock waves melted manufactured/model "meteorites" of various materials and then compared my results with those effects felt by actual meteorites, which experienced similar conditions as they entered the Earth's atmosphere (Zotikov, 1959).

I was unaware at the time that my experiments from the point of view of mathematical physics were exactly similar to those of Dr. Robin. The heat transfer through pressurized, heated air behind a supersonic shock wave through the space between the wave and the solid boundary of the nose cone, was basically similar to the heat transfer through an ice thickness near an ice divide in the Antarctic from the surface to its bedrock as is shown at Figure 2.4.

Dr. Robin and I both assumed the process to be in steady state, and simplified heat transfer equations in both cases also happened to be equal, consisting of two terms, one for the heat conductivity and one for convection.

The first successful Soviet ballistic missile flight, including successful re-entry was in 1956. I gained my Ph.D. degree in 1957. An era of new and advanced ballistic missiles began, with numerous employment opportunities resulting from the growth in this prosperous industry. I became disenchanted, though, partly because of the inherent secrecy of my work and partly due to being part of a crowd of people all doing the same thing. After many years of hard work to receive a Ph.D. in such a demanding branch of science, I had time to look around and discovered that a major scientific event was being planned, the first Soviet Antarctic Expedition (SAE), part of a larger multinational program with an emphasis on the study of the Antarctic continent. The SAE represented the Soviet Union's role in the International Geophysical Year (IGY, 1957–1958). It was the first Russian expedition to Antarctica since 1821.

A friend of mine and co-worker, a mountain climber, was accepted for the expedition as the leader of a group assigned to construct a new scientific station in the middle of the Antarctic Ice Sheet (later named Vostok Station). I was aware that this station would be located at 3,500 m above sea level on a plateau, which is why as a mountaineer my friend was given the job. I pleaded with him to take me along as a

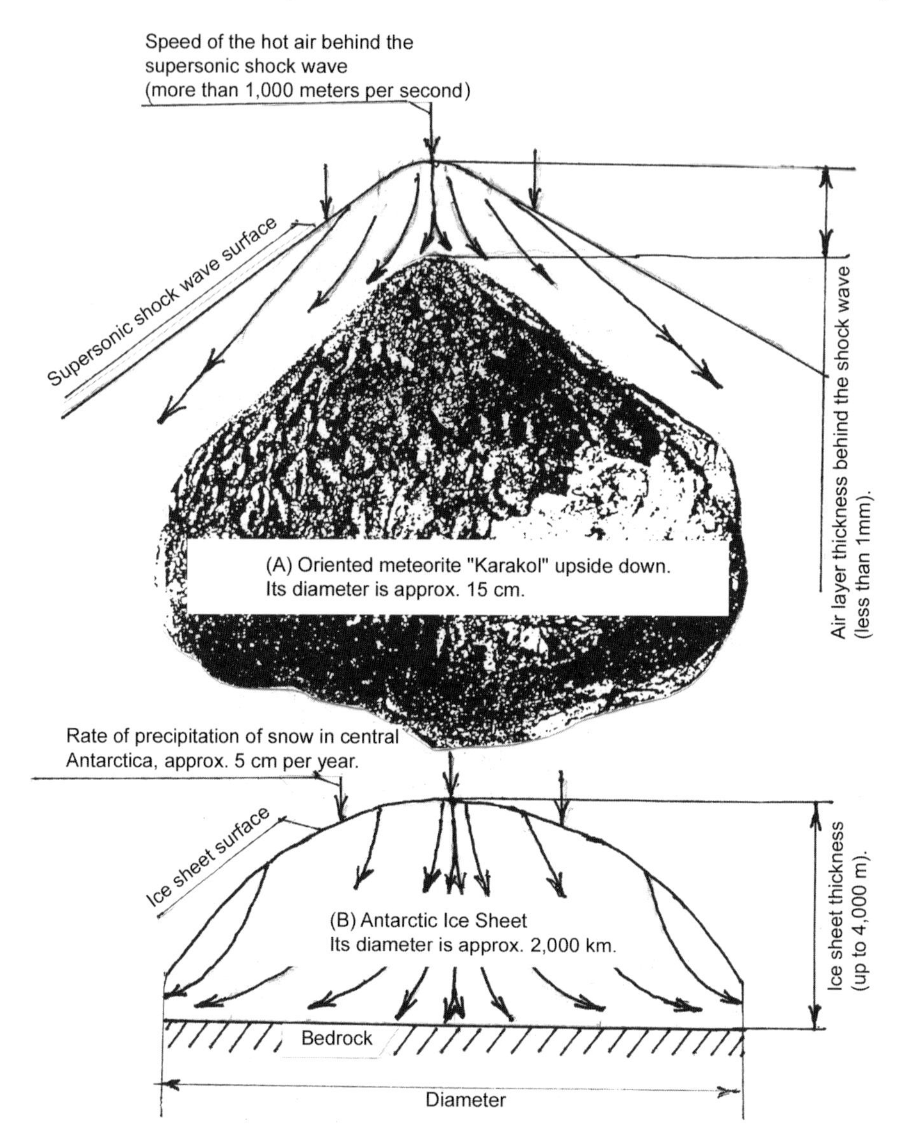

Figure 2.4. An oriented meteorite melting and the Antarctic Ice Sheet bottom melting analogy. (A) Oriented meteorite ("Karakol"). Karakol is a particular name of one of the oriented meteorites, used in the studies of Zotikov (1959). There is a permanent new air supply through the supersonic shock wave surface to the space between the shock wave and the meteorite's surface, determined by the air layer thickness behind the shock wave (order of some parts of a millimeter). There is a permanent meteorite surface melting (evaporation, sublimation) near a front stagnant point (leading nose) area. (B) Antarctic Ice Sheet. There is a permanent new ice supply through the upper ice sheet surface to the space between the ice sheet's surface and its ice–bedrock boundary, determined by the ice sheet thickness (order of some thousands of meters). There is a permanent ice melting at the bottom of the central part of the Antarctic Ice Sheet.

Figure 2.5. Academician Treshnikov was instrumental in constructing Vostok Station just above Lake Vostok.

member of his group, stating that I would be willing to do everything including shoveling snow, washing dishes, any role at all, but he said "No".

His reply in some respect was an act of providence, for later events would provide me with the opportunity to work in Antarctica as a scientist instead of one of many anonymous people who opened Vostok Station, as well as giving me the opportunity to be one of the first to reach the surface of the ice cover of Lake Vostok. As it turned out, my mountaineering friend and his group never reached their destination. They were prevented by many unfortunate circumstances, leaving them about 300 km short of their destination. Instead, a small station (Komsomolskaia) was established, cargo was left there, and the team were flown back to Mirny. Months later, before the end of the expedition, its leader, Dr. A. F. Treshnikov, famous Russian polar explorer and later academician of the U.S.S.R. Academy of Sciences, organized and took command of a new convoy of large tractors. His team drove to the place where Vostok Station was erected at the end of summer 1957. Since then, the base has been occupied nearly every summer and winter. Figure 2.5 shows academician Dr. A. F. Treshnikov.

Deep in my heart I dreamed for the opportunity of being accepted on an Antarctic expedition. I realized now that I must find a different approach, so I applied directly to the very top of the organization in charge of expeditions, and convinced the scientific leader of the new Antarctic expedition that I could apply my knowledge of solving heat transfer problems to the study of the heat regime of the Antarctic Ice Sheet. I was accepted, and at the end of 1958 sailed away to spend a winter at Mirny Station as "glaciologist–thermophysicist" of the 4th SAE.

My meteorite-melting study thus became important for my glaciology applications when I went to Antarctica to study the heat regime of the Antarctic Ice Sheet. The mechanisms and non-dimensional equations of vertical heat transfer through the thick Antarctic Ice Sheet, when applied to a subglacial bedrock surface in its central part near an ice divide, are similar to the equations of heat transfer from a hot supersonic shock wave to a solid surface of a leading nose (a stagnant point) of an oriented meteorite. Especially interesting and unexpected was the situation where non-dimensional Péclet numbers for both cases were more than 1. A Péclet number is a multiplication of a "characteristic" size for a "characteristic" speed divided by a temperature conductivity coefficient. A Péclet number represents the ratio of heat transport by convection to heat transport by conductivity.

The boundary equation of heat transfer between the solid body of my "meteorites" and a liquid (a layer of air between the solid body and shock wave) was also important in the study of the heat regime of the ice sheet. It included the heat flow through the solid to the solid/liquid boundary, and heat flow through the layer. For steady-state conditions, the difference between these flow rates was used for solid surface melting of the leading nose of my "meteorites". This gave me the idea to develop a method to calculate thermal conditions at the bottom of the ice sheet in order to determine whether melting is occurring, using Zubov's "critical thickness" concept, but on a more accurate level, calculating a temperature profile across the ice sheet from a thermal conductivity equation with an ice convection term. I also wanted to calculate the speed of melting, if it existed. The vertical velocity of descending ice compaction for this term for the surface of the ice sheet was assumed to be equal to the surface accumulation ratio (a steady-state approach).

Analyses of the data collected on temperature, ice thickness, and annual snow accumulation collected on the traverse of the ice sheet from Mirny Station to Vostok Station, and on to the South Pole using a thermal conductivity equation with a convection term, increased the critical thickness of the ice (Figure 2.6). Nevertheless, the actual thickness of the ice sheet in all its central regions is much more than the critical thickness (Zotikov, 1961). This means that in this area the bottom temperature of the ice is equal to its melting point, and heat flow from the bottom of the ice sheet to its surface is less than the geothermal heat flow from beneath. It also means that heat flow cannot leave, or be removed from, the ice/rock interface because of the high heat resistance of very thick ice. The difference between these rates of heat flow is used to melt the ice at the bottom of the ice sheet. The melted water fills depressions in the subglacial bed and forms subglacial lakes: part of the water moved to the edges of the ice sheet becoming refrozen. Vostok Station is located in just such an area (Zotikov, 1962).

Temperature profiles from the surface to the bottom of the ice sheet were calculated for Vostok Station and Komsomolskaia Station, and showed that the bottom ice temperature in both cases was equal to the melting point of ice (Figure 2.7).

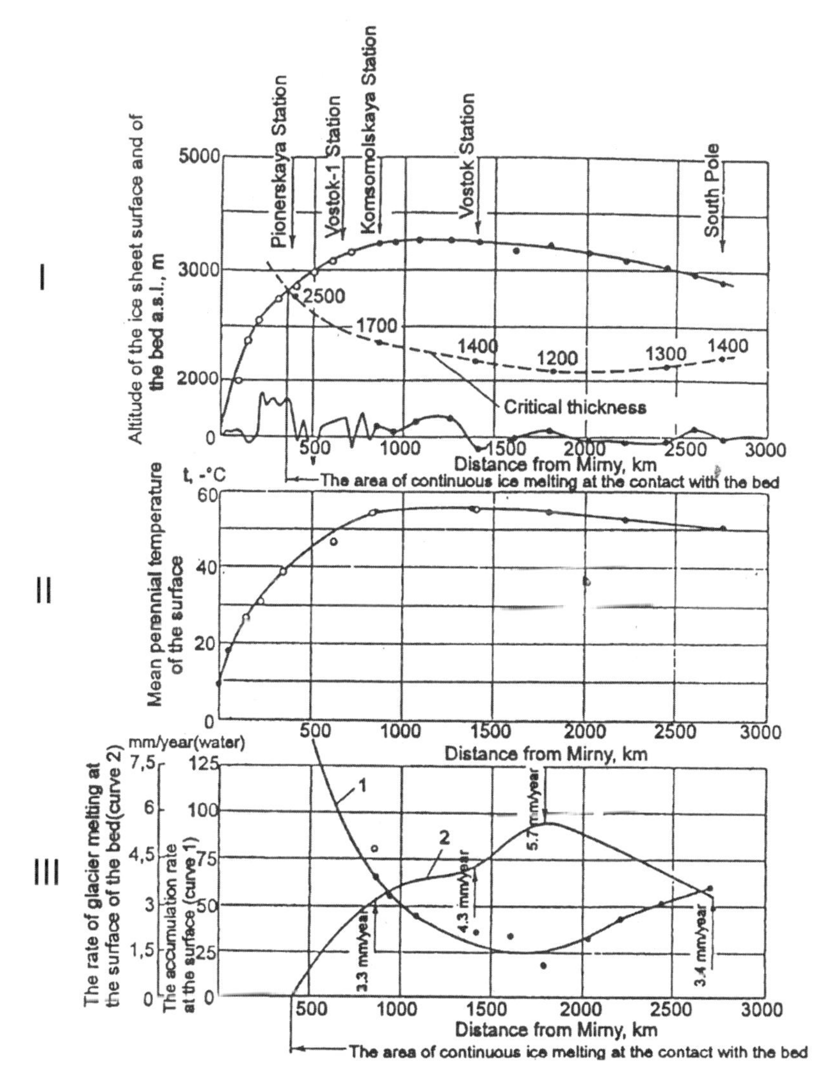

Figure 2.6. Determination of an area of permanent bottom ice melting in the central part of the Antarctic Ice Sheet (Zotikov, 1962). Open circles represent data obtained by the second (1957–1958) and third (1958–1959) SAEs, filled circles represent observations of the fourth (1959–1960) expedition.

These results were quickly recognized by the press. A leading Moscow newspaper *Izvestia* published an interview with me in 1963, the article was entitled "Melts or Does Not Melt". The article stated "In analyzing heat transfer processes within the Antarctic Ice Sheet, our group made an interesting conclusion. The bottom layers of the ice sheet were melting permanently ... ," furthermore, "It is possible to suppose that there is a sea of freshwater there ... that contains oxygen,

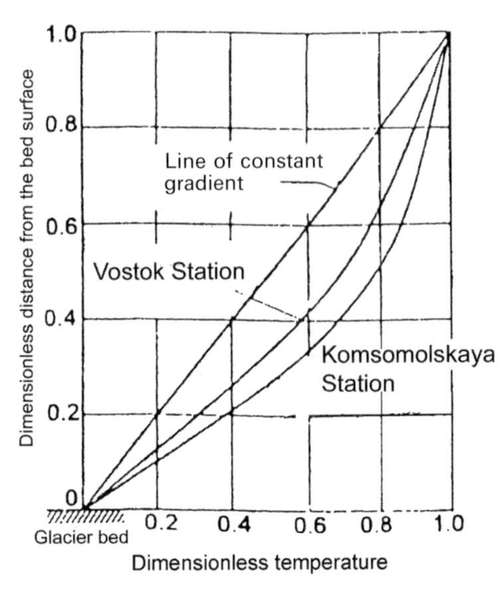

Figure 2.7. Temperature profiles for an ice sheet at Vostok Station and Komsomolskaia Station (Zotikov, 1962). The "0" temperature is the ice melting temperature at the bottom of the ice sheet.

which is supplied to the water from the surface by ice descending from the surface by compaction. It is also possible that there is some unusual type of life there. What kind of life? We do not know..." (*Izvestiya*, 1963).

Our calculations have shown (Zotikov, 1962) that the rate of permanent melting at the bottom was about 1–4 mm of ice per year. This ice formed originally at the surface of the ice sheet as a result of snow precipitation and accumulation. Snowfall at the surface formed a layer that was transformed first into firn and later, by compaction, into dense glacier ice. The measured density of ice at the ice/firn boundary was about $0.70–0.75 \times 10^3\ \text{kg m}^{-3}$, relatively low because the ice has many unconnected air bubbles. The density of pure ice without bubbles is about $0.90–0.95 \times 10^3\ \text{kg m}^{-3}$. This means that each volume of ice that melted at the bottom brought with it at least 15% air under conditions at the surface of the ice sheet. It was easy to calculate how much of the air, under pressure beneath the ice sheet (about 300 bars), was derived from the surface, carried to the bottom, and released into the subglacial water. In this sense the ice sheet works as a huge, high-pressure air compressor! (It was not known at the time that gases at pressures and temperatures present at the bottom of the ice sheet form solid cryohydrates and gas hydrates.) So, I thought (Zotikov, 1977), that this air would be partly dissolved in the water. The time required for this "compressor cycle" was of the order of a million years. This means that the height of a water column brought to the bottom for that time period would be of the order of 0.001 m per year × 1,000,000 years = 1,000 m, and the height of an air column brought to the bottom for this period would be 15% of the height of water, or 150 m for pressure conditions at the surface, or about 0.5 m compressed to 300 bars representing the bottom pressure. This air layer is essentially insignificant and it would be dissolved in the water if we did not consider the following complication to the process.

We mentioned earlier that in our early model the central area of the bottom melting of the ice sheet was surrounded by a belt, where the thickness of the ice was less than the critical thickness. We mentioned that the water from the central area moves to this belt of thinner ice and slowly refreezes (this ice is different from ice formed from accumulated and compacted snowfall). It is known that the process of slow freezing of water is a good means of separating out any inclusions within it. So, it is possible that the water would go to the edges of the ice sheet in the form of a layer of clear ice, free of gases. These layers of clear ice derived from refreezing are commonly observed in icebergs near the coastal perimeter that have become inverted in the water. If this mechanism of refreezing of meltwater occurs, it is reasonable to expect to find a sufficient layer of compressed air above the subglacial lakes of Antarctica, at least in some areas. This possibility presents a useful scenario for writers of science fiction (Figure 1.3).

When I presented these calculations and hypotheses at the meeting of the Soviet Antarctic Committee in 1963, the leader of the third SAE, who led the Russian traverse tractor–sledge party to the Pole of Inaccessibility (Sovetskaia Station), remarked excitedly:

> *Completing the traverse there we decided to blow-up all the unused remaining explosives that we carried for seismic work, about a ton of it. When we fired the shot, we became alarmed because the surface of the ice moved back and forth wildly, and dangerously sounding noises came from beneath. We felt that we were on the top of the roof of a dome, empty inside, and this roof and the dome were strongly damaged by this blow. We felt that we would fall down through to the empty interior . . .*

In contrast to many of the old sailors' stories from remote parts of the world, we can propose a reasonable scientific explanation for this event.

For the moment, though, we considered that the empty dome filled with air under high pressure existed above the subglacial lake. Figure 2.8 shows a photo of Dr. Zotikov when he worked on his subglacial melting ideas.

The next step in developing an understanding of subglacial melting in central Antarctica was presented in an article and a map at the International Symposium of the Scientific Hydrology Association (Zotikov, 1963), which showed that an area of permanent melting at the bottom of the central part of the Antarctic Ice Sheet, with a melting rate of a few millimeters of ice per year, was a large one, covering hundreds of thousands of square kilometers, incorporating the sites of Vostok, Byrd and Amundsen–Scott Stations (Zotikov, 1963). A map of the area is shown in Figure 2.9.

The probability of life existing below the ice sheet was of less interest to people at that time than other aspects of the study (e.g., the existence of highly pressurized air or the influence of bottom melting on the mass balance of the Antarctic Ice Sheet).

Figure 2.8. Dr. Zotikov in 1965 in the old South Pole Station library (at about the time he published his articles about subglacial melting in the central part of the Antarctic Ice Sheet).

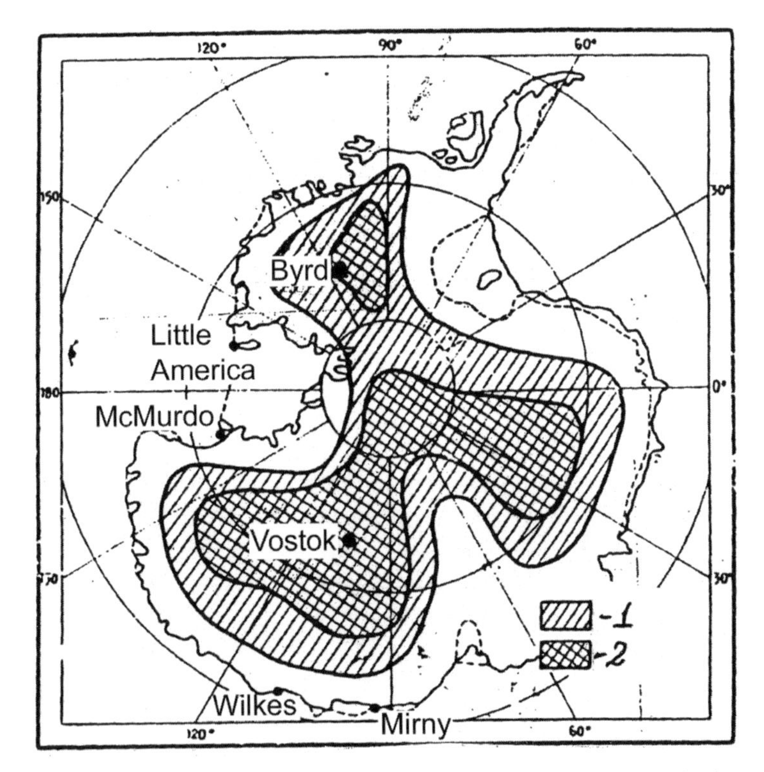

Figure 2.9. Map of Antarctica showing regions of permanent ice melting at the bottom of the central part of the ice sheet (adapted from Zotikov, 1963). 1 – melting region, calculated with an assumption that only geothermal heat flow, averaged for the Earth's surface at 52 $mW\,m^{-2}$, comes to the ice/bedrock interface from below; 2 – a heat flow twice as large comes to the ice/bedrock interface (104 $mW\,m^{-2}$).

The same approach to the probability of the existence of life below the ice sheet was repeated by American scientists when they penetrated the ice sheet and reached the bottom at Byrd Station, West Antarctica, and similarly when the bottom of the ice cover was reached above fast moving ice streams and the presence of liquid water was found in West Antarctica. All these situations are comparable, and can be explained by the same mechanisms.

3

Does subglacial water exist?

An experimental, hypothetical discovery of a subglacial lake under the ice sheet below Vostok Station was accomplished in 1964, and later inferred when direct contact with liquid water was made in 1968 under the ice sheet below Byrd Station.

Very important measurements for understanding the glaciology of the ice sheet in the vicinity of Vostok Station were accomplished by Dr. Kapitsa in 1959 when he conducted the first seismic study of the area using a new technique. On the basis of his previous experience of seismic sounding in Antarctica he found that placing geophones close to the snow surface on the ice sheet was a source of possible errors when it came to interpreting reflected signals due to background noise on the seismograms. To reduce the noise he placed the geophones as deep as possible into the ice sheet, some as deep as 40 m. A special rig for drilling holes for geophones was transported by sledge and tractor to Vostok Station, this led to a significant improvement in the seismograms. Kapitsa (1961) showed in 1958 that the thickness of the ice sheet at Vostok Station was about 3,700 m, not 2,000 m as previously believed. Figure 3.1 is a photograph of Dr. Kapitsa at the time he undertook these soundings.

In a second seismic study of the ice thickness at Vostok Station in 1963, Dr. Kapitsa recorded two reflections from the bottom area. He interpreted the upper reflection as the lower boundary of ice, at a depth of 3,750 m, and the lower one as the lower boundary of a frozen sedimentary layer (Kapitsa, 1968).

It took 30 years for the scientific world to learn that the space between these two reflections was water – Lake Vostok. Until that time the seismogram was never published or made available for study by other scientists. The seismogram was reinterpreted on the occasion of the first International Workshop on the Study of Lake Vostok in 1994 and the upper reflection was identified as the ice/water boundary, the "sediment layer" was identified as the 500-m water layer that is now known as Lake Vostok. This seismogram and the discovery of the thick water layer below the ice sheet at Vostok Station ware presented by Dr. Kapitsa

Figure 3.1. Dr. Kapitsa at the time he worked on reinterpretation of his seismograms, taken at Vostok Station.

at the Scientific Committee on Antarctic Research (SCAR) meeting in Rome in 1994. It was accepted by its delegates with great enthusiasm and was published in 1996 (Kapitsa *et al.*, 1996). The seismogram is shown at Figure 3.2.

I am surprised that neither Dr. Kapitsa nor I attempted to relate this second bottom reflection to the concept of a subglacial lake. We both thought about it, and skirted around the subject at the time, but did not pursue it. Our collaborative work and discussions about a permanent melting at the bottom of the ice sheet below Vostok Station, and the possibility of the existence of a subglacial lake were already published.

In 1963, Dr. Kapitsa and I proposed a project for a Sub-glacial Autonomous Station (SGAS, or PLAS in Russian), for the penetration of the Antarctic Ice Sheet to study the bottom water layer and the possible subglacial lake. The project was based on a small nuclear power plant that would be lowered, as part of a hermetically sealed container of the SGAS, which would contain appropriate instruments and equipment. The nuclear power plant had to produce enough energy to melt the SGAS down to the bottom of the ice sheet. No hole was needed, as the SGAS would sink in a water cavern as the ice melted below it, and the resulting meltwater would refreeze above it. Communication with the ice sheet surface would be maintained by wireless means. This project was approved by the Atomic Energy Institute (now Kurchatov's Institute) of the U.S.S.R. Academy of Sciences at a special meeting, in which the Institute agreed to provide a small nuclear energy reactor of 100 kW, small enough to be installed in the SGAS's 0.9 m diameter cylindrical container. However, this project was never realized. We made the mistake at the beginning

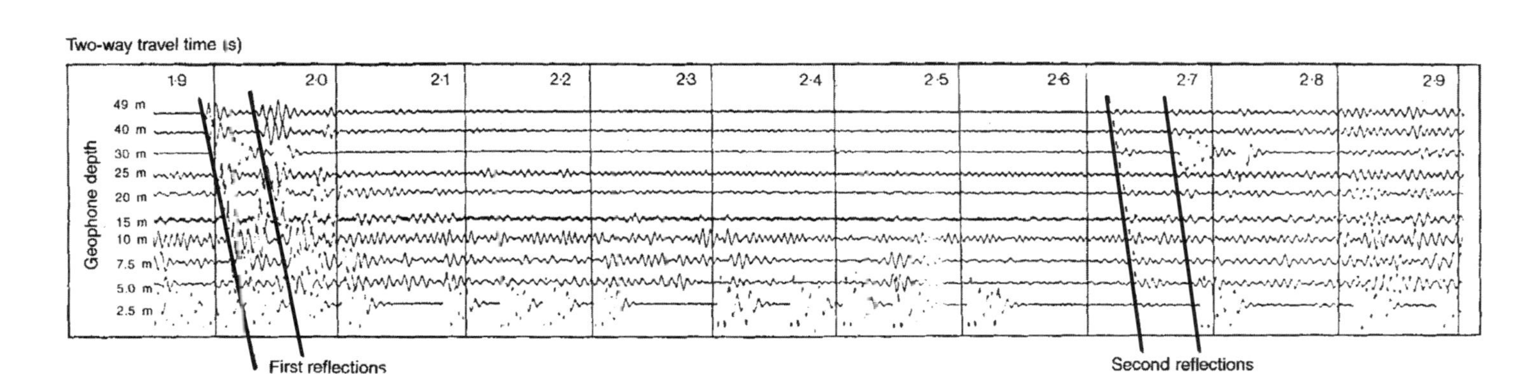

Figure 3.2. Seismogram taken by A. P. Kapitsa at Vostok Station in 1964, reinterpreted by him in 1993 and presented to the SCAR meeting in 1994 (adapted from Kapitsa *et al.*, 1996). This figure represents a section of a 24-channel seismic record with two bottom reflections. The space between the "first reflection" and "second reflection" was interpreted in 1965 as a sedimentary layer (Kapitsa, 1968) – a view accepted until 1993. In 1994 the first reflection of this figure was reinterpreted as an ice–water interface (Kapitsa *et al.*, 1996). The figure records movement over a vertical line of seismometers from 49 m to 2.5 m depth in a borehole situated 180 m from an explosion. It covers the period from 1.85–2.9 s after the explosion of a 5 kg charge of TNT at a depth of 39 m. A conventional horizontal spread of 12 seismometers at 20-m intervals recorded the same echoes, these are not shown. The echo from the bottom of the ice sheet, which is interpreted as an ice–water interface reaches the deepest (49 m) seismometer first at approximately 1.91 s. It then travels up the seismometer line to the surface where it is reflected down to pass the 49-m seismometer approximately 50 ms later. This has a mean velocity of about 2,200 m s^{-1}, typical of compression (P) waves in the top 50 m of firn in this region. About 45 ms after the first arrival, a second wave train of similar intensity and duration follows as result of initial surface reflection of the explosion at 39 m depth. There is no significant return of energy between approximately 2.00 and 2.63 s, when a weaker wave train passes up and down the seismometer line at the same velocity. This confirms that they are compressive (P) waves, the only waves that travel through water, and not transverse (PS) waves, which are sometimes recorded from shots on ice shelves (Kapitsa *et al.*, 1996).

of planning to remain a strictly national project and as such classified (because the nuclear energy reactor was classified). The SGAS was never built. In a comparable example, Philberth (1974) devised a thermoprobe, with an energy supply located at the surface, that melted its way downward through the ice, with the melted water refreezing above the probe. One element that prevented this experiment ever being realized was the nuclear-powered device – something that in the 1960s raised concerns. Additionally, we did not know how to retrieve the SGAS to the surface of the ice sheet. We could not abandon it because the Antarctic Treaty prohibited the disposal of radioactive waste or nuclear power plants in Antarctica. Nevertheless, writing the proposal showed that we were giving serious thought to the existence of a lake below the ice sheet. The lake remained undiscovered for the time being.

The idea of the use of nuclear power to penetrate the ice was discussed again a few years ago. The problem of bringing the SGAS back to surface was solved, at least theoretically: the SGAS was to be designed to float in water – it would be inverted with its hot side up so that it would melt the refrozen ice on the way back up to the surface (Zotikov, 1993).

This idea received new discussion in 1994, after the first Cambridge workshop of 1993, when it became clear that a large subglacial lake below Vostok Station did exist. The same Atomic Energy Institute (now Kurchatov's Institute) agreed to my suggestion for the design of a new and different unit. The institute suggested a unit powered by a nuclear thermoelectric plant that would produce 0.5–2 MW of heat and 10 kW of electricity, with a capability to continue at that level for many years in an environment such as that of Lake Vostok.

However, the general attitude for the use of nuclear energy in the form of a power plant changed, thus making it unlikely that a plant of this type would be acceptable for this application. On the other hand, the concept could be used for making an SGAS device for the study of a subglacial ocean on Europa (a moon of Jupiter), or perhaps the subglacial lakes of Mars (the NASA Cryobot concept). However, it is likely that the idea of using nuclear power to penetrate the ice cover of Europa would meet with strong opposition from environmentalist groups because the nuclear-power device for drilling would first need to be launched into space from Earth.

Returning to the mid-1960s, I spent my second overwinter in Antarctica at McMurdo Station in 1965 as a Russian exchange scientist in the U.S. Antarctic Program. As a result of trying to make a homemade drill to produce a hole at the edge of the Ross Ice Shelf, near McMurdo, I had close contact with the drilling team of the U.S. Army Cold Regions Research and Engineering Laboratory. They had the best drilling equipment in the world at that time, and were intending to drill to the bottom of the ice sheet at Byrd Station to collect ice cores and measure the ice temperature and so on. The thickness of the ice sheet there is more than 2,000 m, and the station is located over the area of bottom melting on my map. We discussed this subject, but I had a feeling they did not take the suggestion of bottom melting seriously.

Two more years elapsed, and on 2 February 1967 a large 87 ft (29 m) long electrothermal drill was installed at Byrd Station to start its objective of reaching

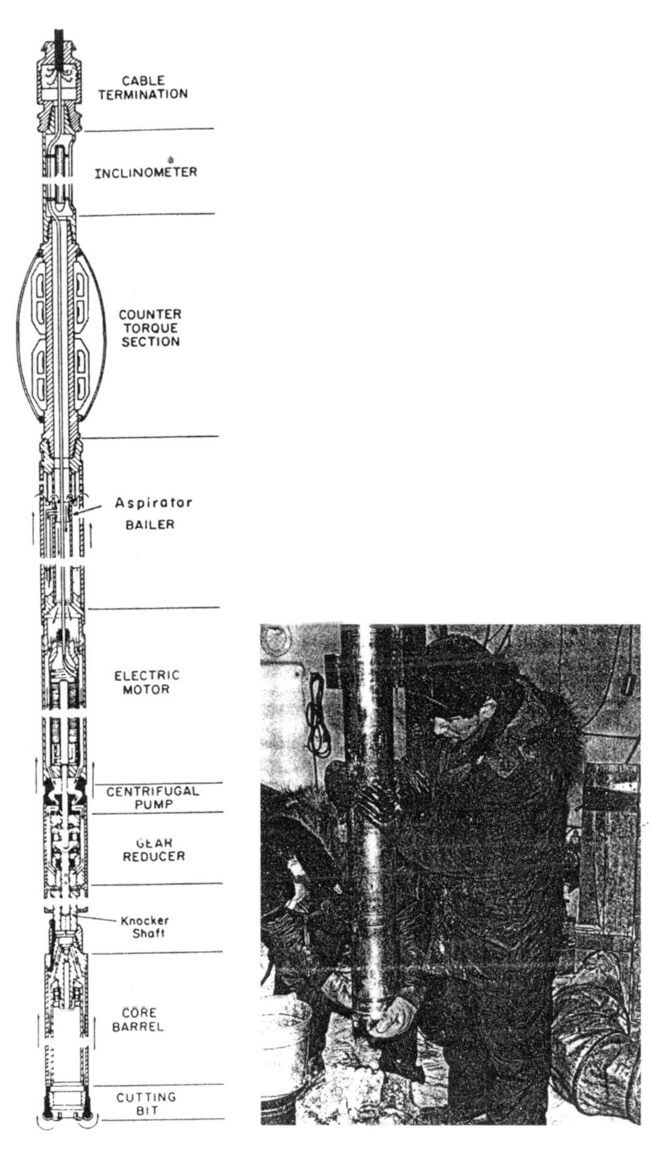

Figure 3.3. American electrical drill unit at Byrd Station (adapted from Ueda and Garfield, 1969). In 1968 this unit at Byrd Station penetrated 2,000 m of ice sheet reaching the bottom and encountering liquid water.

the bottom of the ice sheet – this was the first attempt in Antarctica to reach the bottom of its central region. Figure 3.3 gives some details of this drilling program.

A report by Ueda and Garfield (1969) began:

> *On 29 January 1968 a USA CRREL drilling team successfully penetrated the Antarctic Ice Sheet at Byrd Station after drilling through 7,100 ft (2,367 m) of ice.*

These words were written proudly in the introduction to the report, which went on to describe the problems they encountered whilst drilling. Most of the drilling was accomplished with a cable-suspended electro-mechanical, rotary coring Electrodrill purchased by CRREL from Reda Pump Co., Bartlesville, Oklahoma in 1964. It was invented by the head of the company, Mr. Armais Arutunoff, an oil exploration engineer. After being modified and tested for coring in ice, it was used to penetrate the Greenland Ice Sheet at Camp Century in 1966 (Ueda and Garfield, 1968, 1969). After further modification, it was used during the 1966–1967 and 1967–1968 austral summers at Byrd Station.

It is interesting to note that in the drilling report for Byrd Station, "The primary objectives of the drilling were given: (1) to cut a hole completely through the ice sheet to allow measurements of the temperature profile, the ice flow within the ice sheet, and the ice flow relative to the underlying bed; (2) to provide a continuous, undisturbed core for investigating the physical, structural, and geochemical properties of the ice; and (3) to permit the future *in situ* extraction of entrapped atmospheric gases such as the carbon dioxide used to age date the ice" (Ueda and Garfield, 1969).

There was nothing mentioned in the primary objectives about the possibility (or interest or danger) of encountering liquid water at the base of the ice sheet. The report does, however, mention the incident of encountering water at the base, the original wording of the report being:

> *On 28 January 1968, at a vertical depth of 7,082 ft (2,360.7 m), the first indication of sub-ice debris was recorded ... On the following run, at a depth of 7,101 ft (2,367 m), a sudden decrease in power and a corresponding increase in cable tension was noted, indicating an abrupt change in material had been encountered by the cutting bit. This was later concluded to be, after analyses of subsequent events, a layer of water estimated by the authors to be less than a foot thick. After a few minutes the power increased and drilling was continued to a depth of 7,105 ft (2,368 m). A total of 7.5 ft (2.5 m) of core containing more rock and soil debris was frozen into the upper part of the core barrel. No sub-ice core was recovered.*
>
> *By the time of the next run, which was several hours later due to equipment repairs, it was noted that the fluid in the hole had risen from 630 ft (210 m) to 313 ft. (104.3 m) from the surface. The glycol column down-hole had risen from 5,743 ft (1,914 m) to 5,557 ft (1,852 m) ...*

Some 30 years after this dramatic event, I was drilling a hole in 1977 through 416 m of the Ross Ice Shelf. I encountered a water level, just a few meters before the subglacial ocean water, and experienced events comparable with the drilling at Byrd Station (Zotikov, 1979), indicating that subglacial water had been reached and that the rising liquid level meant that the water would soon encounter the upper, cooler parts of the hole and freeze. Therefore, the message was clear – all equipment must be removed before it became frozen in the hole and therefore damaged, and probably irretrievable. It is easy in hindsight to realize what happened after 30 years of work in this field. In the last days of January 1968 during the drilling project at Byrd Station, further comments from the report are worth repeating:

> *... Various drill surfaces were showing signs of rusting, a phenomenon never noted previously. The only visible evidence of the nature of the sub-ice material were thin films of clay on the drill surfaces ...*
>
> *... The effects of the down-hole water caught in the drill sections which had subsequently frozen during the trip out of the hole began to have serious consequences ... Damage to parts of the drill from the high freezing forces created additional problems and delays. For fear of [...] the loss of the drill, attempts to obtain a sub-ice sample were terminated on 2 February 1968.*

By coincidence, in the beginning of 1968 I completed a draft of my DSc. thesis (Zotikov, 1968), and the permanent bottom melting in the central part of the Antarctic Ice Sheet was one of its main topics. The scheduled day of the defense of my thesis (in Russia we use this term instead of "examination", as in France) before the Scientific Council of Arctic and Antarctic Research Institute in Leningrad was announced as 11 June 1968 – a very important day for me. There were to be many scientists present who did not believe in warm bottom ice and liquid water at the bottom of the central part of the Antarctic Ice Sheet, and I received negative reviews from two leading glaciologists in my country on this matter, which was the main topic of the dissertation. It was mentioned to me that agreement with my idea by the Scientific Council of the Arctic and Antarctic Research Institute was questionable.

Shortly before the date of my "examination", I received an important and timely message from my friends in the U.S. Antarctic drilling team. They informed me that they reached the bottom of the ice sheet at Byrd Station, and unexpectedly had found water.

I saved this telegram for the day of the defense and at the appropriate time the Secretary of the Scientific Council read it. Perhaps because of that everyone in the council voted unanimously in favor of my thesis. Unexpectedly, the uncertainty of the scientific community was so high that it took another two years for the Highest Attestation Committee of the U.S.S.R. to approve the decision of the Scientific Council.

Regarding the drilling operation at Byrd Station, it should be mentioned that very little of the drill could be recovered making it difficult to proceed further with the experiment – so the operation was closed down completely. The manufacturers of the equipment did not get an extended contract, so the experience, skill, and momentum of the best ice cap drilling program in the world was lost.

It took the U.S. Antarctic Program nearly 15 years to recover and catch up with international levels of deep-ice drilling.

I seldom read anything that mentioned the possible existence of liquid water below the ice sheet under Byrd Station; similarly the literature did not contain information about the necessary precautions required to prevent water from getting into and rising up the drilling hole. For example the increase in the height of a column of unfrozen drilling fluid, which was put into the hole to compensate for the Mountain Pressure[1] and prevent hole closure, could prevent the sequence of events that would close the hole and freeze around the drilling equipment. In addition, no one anticipated the possibility of organisms in the water until 20 years later.

[1] The Mountain Pressure at any depth is jargon from Russian mining engineers, meaning the hydrostatic pressure equal to that of a column of rock (or ice, in this case) above this depth.

4

Discovery of subglacial lakes by radio-echo sounding

Nearly 10 years passed after the first publications appeared on the theory of subglacial melting and the existence of lakes in central Antarctica, before a new related development occurred. It started with the use of radio-echo sounding altimetry in aircraft over Antarctica. Pilots of these aircraft reported that the altimeters occasionally showed erroneous altitudes when compared with actual altitudes when their planes flew low above a glacier or an iceberg. This difference was explained as a result of a mixture of a radio-echo reflection from an air/surface boundary and a reflection from the bottom of the glacier or from internal layers within the glacier. This discovery implied that finding the proper frequency and developing appropriate monitoring equipment would result in radio-echo sounding as a new and powerful method for the study of ice sheets.

The first experimental discovery using this method included a series of subglacial lakes below the central part of the Antarctic Ice Sheet by a group of British scientists led by Dr. G. de Q. Robin, at that time the Director of the Scott Polar Research Institute (SPRI) of Cambridge University, and author of the article on the heat regime of the Antarctic and Greenland Ice Sheets (Robin, 1955). The SPRI group worked in close collaboration with the Electromagnetic Laboratory of the Technical University of Denmark (TUD) and the Division of Polar Programs of the U.S. National Science Foundation (NSF). This cooperation was advantageous because it combined the expertise in electronics and antenna design at the TUD, developments of new equipment at the SPRI, and the logistical capabilities of the American LC-130 aircraft for installation of the equipment and long-range flights over large areas of the ice sheet with their highly accurate inertial navigation systems (Robin *et al.*, 1977). These were the days before global positioning systems (GPS or GLONASS) were in use, when flying over featureless terrain was difficult, particularly if you required accurate navigation. As a result, many of the flights, which are shown in Figure 4.1, had to be made within viewing distance of Vostok Station, thus ensuring a reduction in navigation errors.

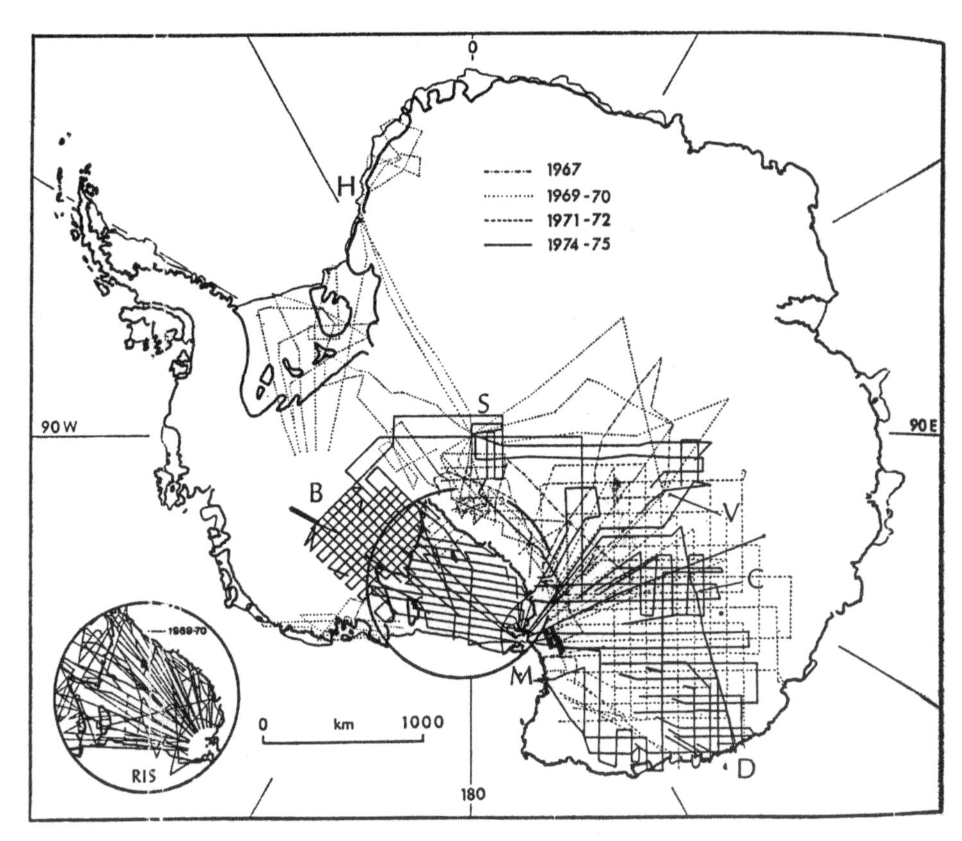

Figure 4.1. Radio-echo sounding flight lines in Antarctica undertaken during 1967–1975 operations (adapted from Robin *et al.*, 1977). Abbreviations: RIS – Ross Ice Shelf; Antarctic stations: B – Byrd; C – Dome C; D – Dumont d'Urville; H – Halley Bay; M – McMurdo; S – South Pole; V – Vostok.

Most of the results from these flights consisted of lengthy film records, during flight paths, of time versus positions of radio reflections taken by continuous monitoring from within the ice sheet. The film records show the ice/rock interface (the bottom of the ice sheet) as well as numerous near-parallel, internal layer reflections. The bottom of the ice sheet in the majority of the reflections appears irregular with a vertical roughness resolution of more than 500 m and a horizontal length of about 300 km.

In some places the appearance and strength of the reflections were completely different, appearing as horizontal and smooth. An area with especially strong reflections of this kind was recorded near Sovetskaia Station in the central part of East Antarctica, where the ice thickness was about 4,200 m. This was explained by Dr. Robin and his colleagues as an ice–water layer boundary.

Further studies, conducted during the 1971/1973 field seasons, showed that the flat, smooth surfaces (some kilometers in length) with a high reflection coefficient are common elsewhere in East Antarctica (Oswald and Robin, 1973).

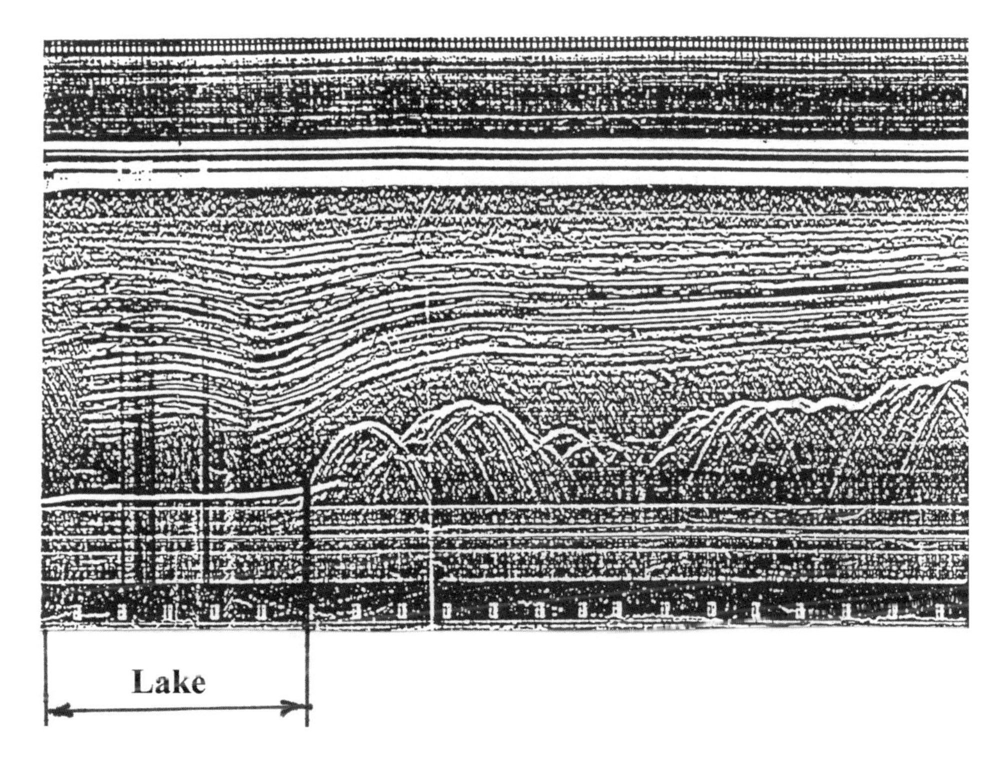

Figure 4.2. Radio-echo sounding showing a subglacial lake (G. de Q. Robin, 1993, pers. commun.).

Studies on these peculiarities of bottom reflections were continued in the 1974/1975 field season. Analyses of 17 other cases of this kind in East Antarctica showed that all were located in areas with evidence of an ice bed composed of solid rock, where the rate of glacier movement was low (meaning that transport of erosional debris was also low), and the surface above those areas was also nearly horizontal. Dr. Robin and his colleagues interpreted this phenomenon at the bottom of the ice sheet as lakes (Figure 4.2).

Unfortunately, the thickness, or depth, of the lakes was unknown because radio signals do not propagate in water. It was evident only that the thickness is large enough to be larger than the radio-echo signal's wavelength.

Many radio-sounding flights were made during the period 1971/1972 and 1974/1975 in the vicinity of Vostok Station, because the majority needed visual contact with the station in order to correct for aircraft navigation. A number of those flights, which registered a large number of "water reflections", led Robin *et al.* (1977) to suggest that there was a large subglacial lake in the area, with its center located about 190 km to the north-northwest of Vostok Station. So, the discovery of Lake Vostok by radio-echo sounding had been made, but at that time nobody was aware of the ramifications of their findings.

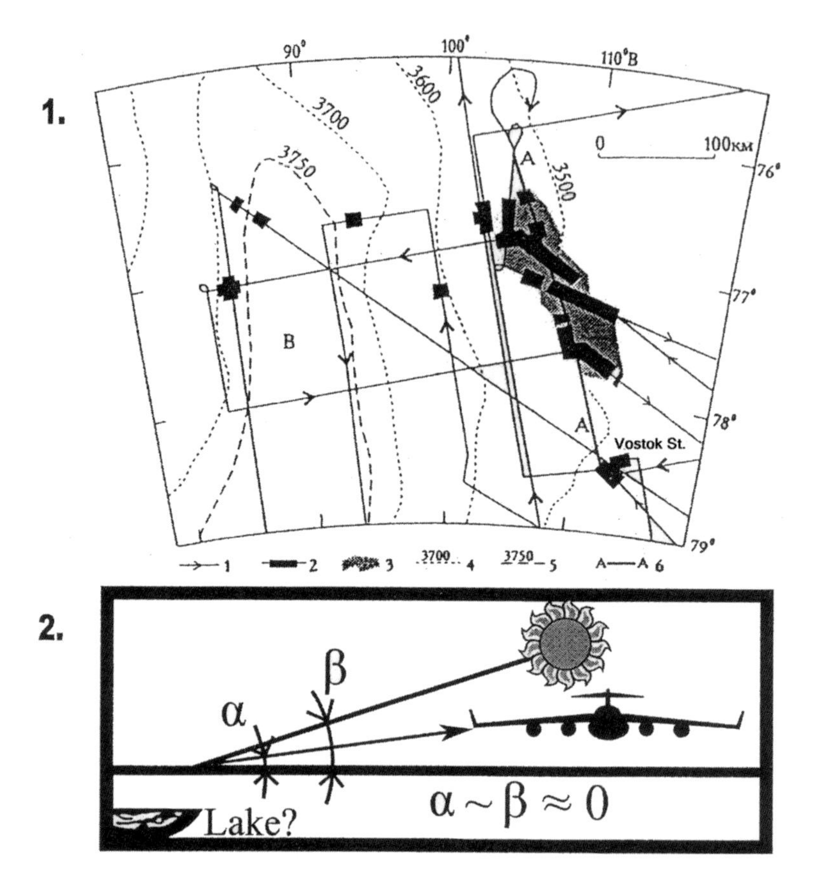

Figure 4.3. A map of Dr. Robin's flights over Vostok Station (adapted from Robin *et al.*, 1977). (1) Sketch map of the location of sub-ice lakes detected by radio-echo sounding at the Dome ("B")–Vostok ("V") region (adapted from Robin *et al.*, 1977). Flight tracks for 1971–1972 and 1974–1975 seasons are shown. Thickened portions of the lines locate sub-ice water bodies. The contours represent heights in meters above sea level. (2) Required angles between the Sun and low-flying aircraft (Robinson, 1960).

To review the situation around Vostok Station: it was known that most radio-echo sounding flights began and ended around Vostok, with each flight identifying the presence of liquid water at the base of the ice sheet. However, a seismic study of the bottom of the ice sheet at Vostok Station was conducted in 1964, with the interpretation that it represented a layer of frozen sediments, not water. This disagreement raised some doubt in the interpretation made by Robin's group, because evidence from seismic studies was traditionally accepted as more accurate than radio-echo sounding.

Despite this disagreement, Dr. Robin insisted on the existence of a large lake, directing special attention to data from flight number 130 along flight line "A–A" in Figures 4.3 and 8.1. He believed that this flight traveled along the long axes of the

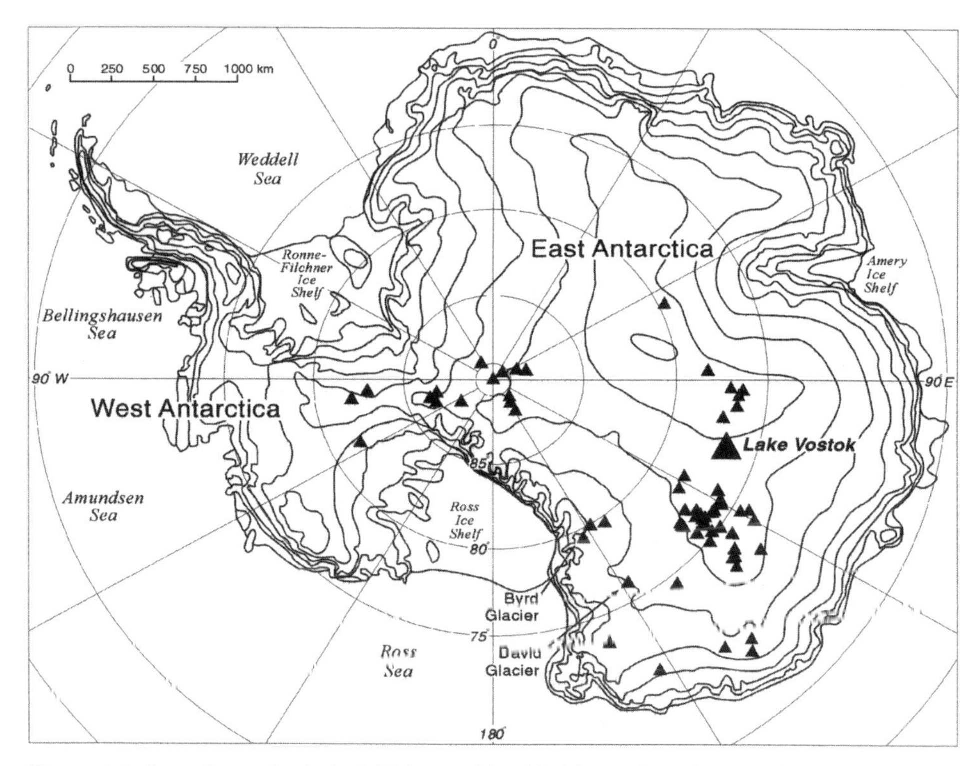

Figure 4.4. Locations of subglacial lakes as identified by radio-echo sounding (adapted from Siegert, 2001). The lakes are concentrated near ice divides.

lake, and crossed the middle of it. The seismic sounding at Vostok Station, however, he believed to be located away from the lake.

It should be noted that the main radio-echo sounding continuous recording camera had jammed film shortly after the aircraft arrived over the lake, and for a major part of the time that the plane flew along the lake it was not working. The scientists on the flight repaired it, and the camera recorded the bottom water reflection again just before the lake ended. The situation was a good example of Murphy's Law. Dr. Robin described the problem: "... one of the radio-echo recording cameras jammed during part of this flight, but the impression formed at the time was that the lake's echoes were present over much more of the discontinuous section A–A [shown in Figure 4.3]. Careful inspection of the 1971–1972 records also revealed lake echoes on flight paths close to this area. We believe that all these echoes originate from a practically continuous body of water, the long dimension of the lake being about 180 km with a typical width of 45 km" (Robin *et al.*, 1977). One could finally say that this was the time of the actual discovery of Lake Vostok.

The operational base for the airborne radio-echo sounding undertaken by Dr. Robin was McMurdo Station, and I was fortunate to be there after I had returned from field work on the Ross Ice Shelf Project (RISP). Dr. Robin told me about his discovery at this time, and I mentioned my idea of detecting traces of large subglacial lakes on the surface of the ice sheet, partly from a visual study of the ice sheet surface from low-flying aircraft. I told him that I spent my first winter in Antarctica with a senior navigator of an aviation group of the Soviet Antarctic Expedition (SAE) of 1958, Mr. Robinson – who had told me that when flying from Mirny to Vostok Station he, and other pilots, had seen some relatively large areas on the surface of the plateau that were distinctly different from the rest of it. These areas were always seen in the same places, and pilots used them for navigation, calling them "lakes" (Robinson, 1960).

However, the "lakes" could be seen only when the low-flying plane was traveling away from the "lake" and its angle of view was low. In a short letter to the editor of the SAE Information Bulletin, Robinson mentioned that "... Natural landmarks in the interior of the continent include, in addition to individual mountains and mountain ridges, oval depressions with gentle 'shores' which are visible from an airplane over the plateau. The depth of these depressions usually does not exceed 20–30 m and their length, 10–12 km. These unusual depressions are sometimes called 'lakes' by pilots. These lakes are clearly distinguishable from the air as spots against the white background of the plateau, especially when the angles of the course of the Sun are close to 180°" (Robinson, 1960). I disagreed with his comment that the "lakes" appeared as depressions with depths not exceeding 20–30 m because no one had yet visited these "lakes" on the ground. The words "depressions" and "depths" may have been arbitrarily added to the article, possibly even at the insistance of the editor prior to publishing the letter. From what Robinson told me I got the impression that the differences at the edges of the "lake" and the "lake" itself were a result of optical properties, hence you could not actually "see" the "lake" (it depended on a low Sun angle, for one thing).

Mr. Robinson and I spent a winter in Antarctica, giving me more time to find out about these "lakes". Soon after his short article was published he regrettably lost his life in an airplane crash in the Arctic.

Robinson's "lakes" at the surface of the central, and hence thickest, part of the Antarctic Ice Sheet, together with Dr. Robin's discovery of a large subglacial lake near Vostok Station, represented a clear, simple, and important message. A subglacial lake, one to two orders of magnitude larger than the thickness of the ice sheet above it, has to be transparent and exhibit a surface expression of its existence. Conditions of an ice sheet moving along the rough slope of a rocky glacier bed and those floating on the water of a sub-ice lake are vastly different. This difference is revealed at the surface, if the size of the lake is large enough compared with the thickness of the ice sheet above it.

It was strange that this thought did not occur to Mr. Robinson and me when we discussed his "lakes" in 1959, because at that time I already knew about permanent melting at the bottom of the ice sheet in central Antarctica.

So, we left Antarctica, and, unfortunately, I never asked him for a look at his flight maps to see where the lakes were located, thinking that we would have sufficient time for that later. Soon after, he lost his life. I later tried to find evidence of these lakes from other pilots, but no one had anything certain to say about them. It looked like Mr. Robinson kept all the information to himself. He was a graduate of the Faculty of Polar Countries of the Geography Branch of the Moscow State University, and hoped to study for a Ph.D. on the subject. In 1975 at McMurdo, in memory of my friend, I tried to find surface evidence of his "lakes" above the huge subglacial lake discovered by Dr. Robin and his team.

Dr. Robin took to the idea enthusiastically, and we soon made a special flight in search of visible evidence. I say "special", because to simulate the conditions of Mr. Robinson's flights the LC-130 aircraft had to fly only a few hundred meters above the surface of the plateau, a practise that was both excessively fuel consuming and dangerous.

Dr. Robin wrote the following about the event:

> *In the course of an informal discussion of this feature ... I. A. Zotikov ... drew attention to a report by Robinson (1960) ... on the existence of shallow snow surface depressions in the Vostok area ...*
>
> *A subsequent flight over the large subglacial lake did in fact give visual confirmation of this, with the 'lake's shores' showing up as areas of whiter snow – corresponding well in some cases with the edges of the radio-echo results. We suggest that these 'lakes' are visible because of the difference in appearance of a uniform flat surface (the 'lake') and the gently sloping one (the 'shore'). This could be due either to changes in albedo caused by differences in relative sun angle between the two surfaces, or less directly to changes of texture resulting from different snow accumulation rates.*
>
> *The extent of the subglacial lake cannot be unambiguously defined, since unfortunately one of the radio-echo recording cameras jammed during part of this flight, but the impression formed at the time was that lake echoes were present over much more of the discontinuous section A–A [Figure 4.3] ... We believe that all these echoes originate from a practically continuous body of water, the long dimension of the lake being about 180 km with a typical width 45 km.*
>
> *The removal of basal friction over such a large area must be expected to produce some effect on the upper surface of the ice sheet, and in our case of a substantially level basal surface, a level upper surface seems a plausible result. Our measurements are not sufficiently precise, but indicate a mean surface slope of less then 1 in 2,000 compared with the regional value of about 1 in 700.*
>
> Robin *et al.* (1977)

A large area where traces corresponding to "water" at the bottom are located and where Robinson's "lake" effect has been seen from low-flying aircraft (Figure 4.3(1)) was shaded by Dr. Robin on the map. We know now that this shaded area is where Lake Vostok is located. The actual discovery of what is now called Lake Vostok was published by Robin *et al.* (1977) in the prestigious journal *Philosophical Transactions of the Royal Society of London*, and although it was seen by many, at the time it did not produce as much interest in the scientific world as it does now.

Some years later, Dr. Robin's student, Mr. McIntyre, compiled a map of East Antarctica with an indication of a large subglacial lake, just where Lake Vostok is located (McIntyre, 1983; Siegert *et al.*, 1996). This lake was marked on McIntyre's map as Lake Vostok, 11 years before the first mention of the name in any other published literature.

5

The need for reinterpretation of seismic data

Part of the ERS-1 satellite mission included a precise laser altimetry of the Earth's surface. The orbital inclination of the satellite covered a sufficient part of the Antarctic Ice Sheet. Ridley *et al.* (1993) collated the laser altimetry data on the ice sheet's surface and mathematically summarized it from many different orbits. This gave them the ability to plot a very precise map of the Antarctic Ice Sheet's surface with an accuracy of 0.5 m within an area of $10\,km^2$. Equal elevation lines on this map showed without doubt that there is a flat, nearly horizontal surface of a floating ice cover just above the area marked by Dr. Robin's bottom water reflections of 1971–1972 and 1974–1975. This ice cover differs distinctly from the slopes of a grounded ice sheet. The distinct shores of the lake are marked by a line of small depressions at the upward side and lines of gentle hills at the downward "banks" of the lake.

As a supplement to his work, Ridley *et al.* (1993) took the map of equal elevation lines for Antarctica and computed an isometric view, imitating a side look from a low-flying aircraft, with shading corresponding to the actual position of the low-angle Sun. It was suddenly clear (Figure 5.1) that there is a large oval-shaped lake more than 200 km long and about 50 km wide near Vostok Station, just south–south-west of it.

At about this time the Royal Society in London and the U.S.S.R. Academy of Sciences signed an agreement to start a program of grant awards. On the basis of this agreement, a Soviet scientist could apply for a grant from the Royal Society to cover expenses in the U.K. for a specified period of time on the stipulation that a prominent member of the Scott Polar Research Institute (SPRI) or any other British Institute would offer an invitation. I expressed my interest to Dr. Robin to work at SPRI on subglacial lake issues, and he replied positively, authorizing a 6-month stay.

The combination of Ridley's article and his map showing a distinct oval shape suggesting a "lake", my arrival at SPRI, and accumulated data provided a good

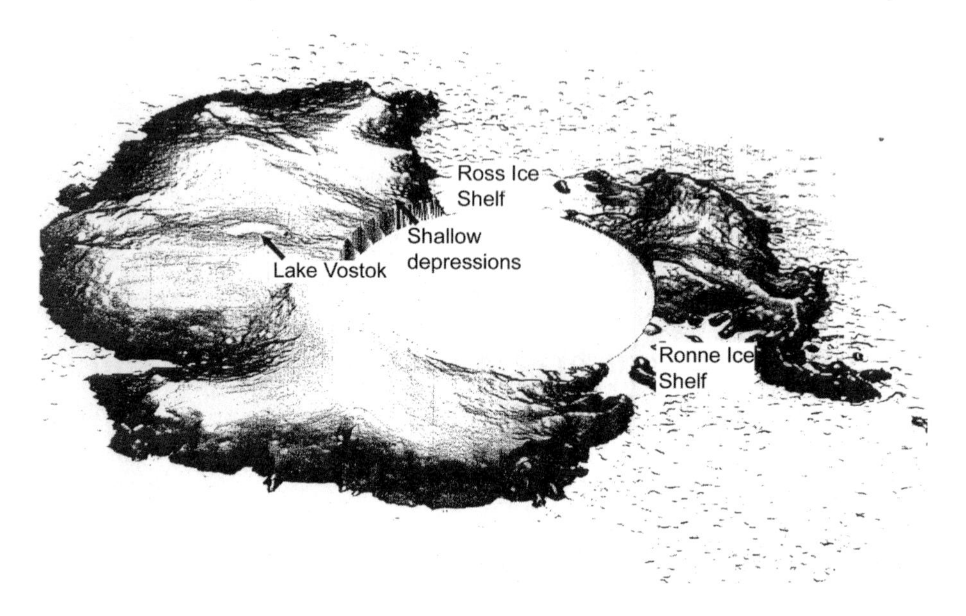

Figure 5.1. An isometric plot of Antarctica compiled from the ERS-1 satellite radar-altimeter data. Details of the surface topography are highlighted, including Lake Vostok (adapted from Ridley *et al.*, 1993).

reason for Dr. Robin to organize a one-day workshop in Cambridge to discuss the scientific problems associated with the existence of the lake. There remained doubts as to whether the lake was real (i.e., whether it represented a reasonably deep layer of liquid water). Some related questions needed to be addressed. American scientists working on ice streams in West Antarctica found them to resemble "rivers" of relatively fast-moving ice compared with the rate of flow of the surrounding ice sheet. Reflections obtained by radio-echo sounding of the bottom of some of these ice streams showed flat features that resembled water. However, deep drilling to the bottom showed no open water existed at these sites, but instead they found a layer of till, fine moraine-type particles saturated with water melted at the bottom as a result of heat produced by fast-moving ice streams (Blankenship *et al.*, 1986).

The lake below Vostok Station was located more than 100 km from the ice divide, with ice arriving at the lake at a rate of about $3\,\mathrm{m\,yr^{-1}}$ (an estimate arrived at from simple calculations). Experimental measurements at Vostok Station gave a similar rate. This is not particularly fast, but it means that the ice sheet continuously produces some kind of morainal debris at the ice/rock interface, which becomes trapped by moving ice, is brought to the lake, and is deposited (Figure 5.2). There should be no question that the ice sheet would have been the cause of this activity – the big unknown was, "How much material had been transported and for how long a time period?"

In preparation for the one-day workshop, we estimated how deep the lake water would be if the glacier did not bring debris into the lake. We took reels of bottom

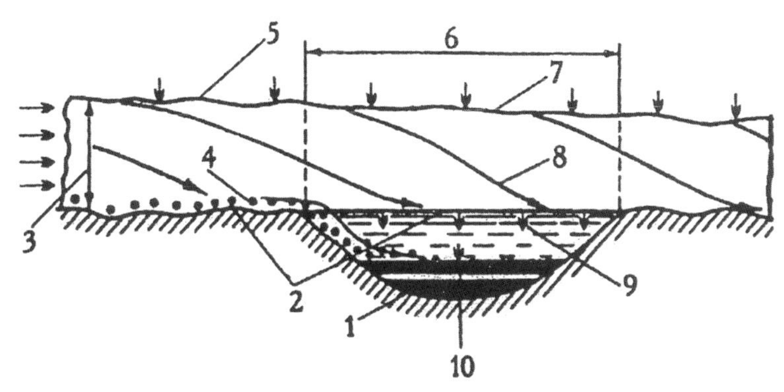

Figure 5.2. Schematic representation of a subglacial lake in the central part of a thick, Antarctic-type ice sheet (adapted from Zotikov, 1986). (1) Depression in subglacial bedrock (lake reservoir); (2) bedrock; (3) thickness of ice sheet; (4) particles carried along by the ice sheet near the bedrock interface; (5) upper surface of the ice sheet; (6) "lake" boundary projected onto the glacier surface; (7) upper surface of the ice sheet above the lake (see Chapter 4); (8) ice flowlines; (9) subglacial lake; and (10) sedimentary layer at the lake bottom

reflections of Dr. Robin's flights from the 1971–1972 and 1974–1975 periods of sounding. The average roughness of a "dry" glacier bottom of East Antarctica for a 200-km section of ice (the approximate length of the lake) in terms of the elevation difference of the highest and lowest parts of the lake bottom was estimated to be about 500 m.

Other studies have shown that the ice sheet moving across the lake could bring about 100 m of debris at the upward bank of the lake with less at the opposite bank. Although the estimates are very rough, they indicate that the lake is probably not filled completely, leaving space for liquid water. These ideas formed the level of our understanding for our "Workshop on Geophysical Studies of 'Lake Vostok'". This was also the first time that the lake was given that particular name – it was a convenient way to refer to it. No one objected to this name from the collective scientific body familiar with the "lake", and it was also considered appropriate because the lake was linked with Vostok Station. I also considered the name "Lake Robinson", after the pilot who discovered the surface indication of the lake and who died soon after while on another polar assignment. His father was English and his mother Russian, and in some way he symbolized close British–Russian cooperation regarding the study of the lake. Dr. Robin did not agree, however, and convinced me that "Lake Vostok" was more appropriate. As of the end of 2004 the name had not been approved by any official geographic body.

From the workshop report that was authored by Dr. Robin and me, the following statements describe Lake Vostok.

> *The workshop reviewed existing knowledge and suggested further studies to enlarge our understanding of 'the remotest lake on Earth' ... The lake has an area of around 10,000 km^2, about one-third the area of Lake Baikal in Asia. It*

lies beneath about 4 km of ice to the north of Vostok Station. The depth of water layer is unknown but is likely to average about some tens to a few hundred meters.

The workshop was organized by the SPRI of Cambridge University and held at a building of the British Antarctic Survey (BAS). Attendance at the workshop included the following: G. Robin and P. Clarkson (SPRI); I. Zotikov and A. Kapitsa (Russians working at SPRI); J. Ridley (Millard Space Laboratory); and C. Doake, D. Vaughan, B. Storey, A. M. Smith, E. King, and D. J. Drewry (British Antarctic Survey). The agenda for the workshop contained five major items: (1) data review; (2) geophysical interpretation of data; (3) future programs; (4) long-term objectives; and (5) conclusions.

Summaries of these items are as follows:

(1) *Data review*. Soviet aircraft navigators first reported in 1959 that oval depressions with gently inclined boundaries, or "shores", were among the natural landmarks on the ice sheet, although they were not located on maps (Robinson, 1960). Extensive studies of the sub-ice relief by airborne radio-echo soundings made by the National Science Foundation–Scott Polar Research Institute–Technical University of Denmark (NSF–SPRI–TUD) program found occasional strong basal reflections over a few kilometers of the flight lines. These indicated specular reflection from a sub-ice–water interface, with a water depth of not less than one meter. The first was reported near Sovetskaia Station in the 1967–1968 season. Seventeen "sub-ice lakes" were reported after the 1971–1972 season when their distributions and other properties were discussed in more detail (Oswald and Robin, 1973). During the 1974–1975 season, similar reflections, shown as heavy black bands in Figure 5.3, were noted along flight lines. The presence of a large "sub-ice lake" was also reported in Robin *et al.* (1977). The overlying ice thicknesses ranged from 3,600–4,200 m. The relatively level ice surface over the sub-ice lake was noted. Digital data on the ice thickness and surface elevations from all NSF–SPRI–TUD flights are now stored on computer at the BAS. Copies of any of the relevant sections are supplied on request by D. Vaughan (BAS), with the original film records on flight data archived at the SPRI.

(2) *Geophysical interpretation of data*. Survey mapping of the surface elevation of the Antarctic Ice Sheet by satellite altimetry from ERS-1 has provided detailed information of surface contours as far south as 82°S. Initial mapping by preliminary or "fast-delivery" data, was used to provide the surface contours in Figure 5.3. About 2,700 surface elevations were used, with mean errors of ice sheet elevations at about 2 m. These errors drop to 0.2 m over the flat central area of the lake. The errors relate to local topography. In addition, satellite orbit errors could be plus or minus 15 m. The most interesting discussion at the workshop was on the possible depth of water in the lake. All our data from theoretical estimates and Robin's radio-echo sounding indicate that there could be a real lake there, representing one of the great geographical discoveries of the

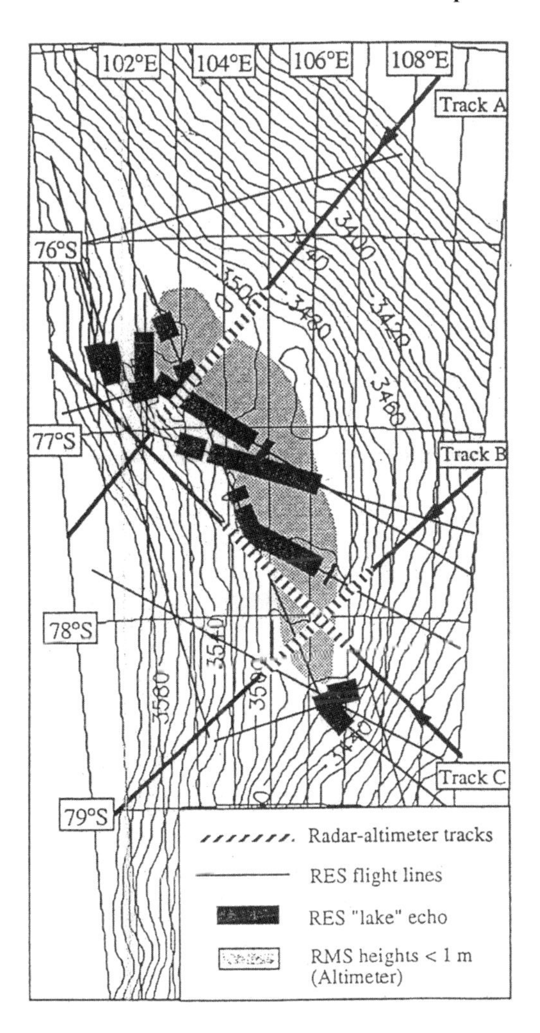

Figure 5.3. Map showing the region of Lake Vostok as it was seen in 1993. Surface elevation lines are reconstructed by Ridley *et al.* (1993) on the basis of low bit rate (LBR) data from the ERS-1 satellite. Also shown are the radio-echo sounding (RES) flight paths and the location of strong bottom "lake-like" echoes from the radio-echo sounding by Robin *et al.* (1977) (adapted from Ridley *et al.*, 1993).

end of the 20th century. However, although Robin's data show a lake under Vostok Station, Kapitsa's seismic data shows no lake – with more confidence attributed to seismic data than radio-echo sounding data. Skeptics wondered if seismic data for other regions, which Robin interpreted as lakes from his radio-echo sounding, would be negative. This uncertainty was also expressed by the editors of *Nature*, reflected in their rejection of a manuscript by Robin in which he first tried to publish a paper on the discovery of a lake using the results of

Ridley's glacial erosion/deposition estimates. *Nature*'s editors said they had already published a paper on subglacial lakes by the same author several years previously, and they would publish new material only if it included at least one independent piece of evidence of the lake's existence (e.g., water thickness data from seismic work). However, this additional evidence appeared to be unobtainable. The answer appeared to lie with seismic work at Vostok Station by Kapitsa (1968), as outlined in the Workshop report:

> *Measurements of ice thickness by seismic reflection shooting were made at Vostok Station and three locations along and near the west edge of the lake by Kapitsa (1968). He interpreted a second, deeper echo at Vostok as coming from sedimentary layering, but the present information suggests that it could come from the base of a water layer around 70 m thick. Over some 50 km from where the other soundings were obtained, the snow surface was extremely soft and made travel difficult. This could result from the flat surface causing snow deposition in light air in contrast to almost continuous winds over sloping surfaces.*

Kapitsa was present at the workshop and agreed with our interpretation and was eager to check the data, resulting in the following statement:

> *... but the present information suggests that it could come from the base of the water layer around 70 m thick. This record and that of the three other soundings near the lake margin will be re-examined in Moscow.*

(3) *Future programs.* The first priorities for future study are as follows: (a) it was expected that assessment of the more accurate data would be completed within a month or two; (b) radio-echo data should be reassessed, especially in regard to their accurate location and the surrounding topography; (c) seismic soundings in the region of the lake would be re-examined by Dr. Kapitsa and his staff in Moscow.

Second priorities included obtaining additional field data to be by standard methods during the next few years. (a) Determination of surface movement by satellite position monitoring at selected points with equipment placed by surface parties or dropped from aircraft. Alternatively, if the recently developed new technique of radar (synthetic aperture radar, SAR) interferometry from ERS-1 could be used to measure velocities, as on the Rutford Ice Stream, this would provide valuable and comprehensive data at a limited cost. Dr. Doake (BAS) was asked to see if this was possible. (b) Ice thickness determination: additional airborne radio-echo sounding is needed to provide better definition of the basal topography over the lake and surrounding bedrock. This should be undertaken as soon as possible. It should also be used to determine whether the small lake segments near Vostok (Figure 5.3) are continuous with the large lake. (c) Water layer thickness: seismic reflection shooting measurements appear to be the best method available. Wide-angle reflection measurements may also provide information on the nature and thickness of basal sediments and bedrock. (d) Long-term monitoring from within the present hole at Vostok. When access to the hole is possible, its use for long-term monitoring (10–100 years) of geothermal heat flow and of microseismic and acoustic activity was suggested. This is of increasing value when the base of the borehole is approaching a sub-ice lake.

(4) *Long-term objectives*. The major long-term objective is to study the geophysics and biology of the lake system. While remote-sensing techniques already discussed can advance a general knowledge of the lake, direct sampling of chemical and biological properties and continuous monitoring of lake waters by instruments *in situ* must be a long-term objective. Extensive discussion of the best techniques of sampling is needed. Particular attention must be given to avoiding contamination of lake waters through the introduction of foreign material such as drilling fluids used in deep boring. Before the existing borehole at Vostok can be deepened to the lake level, a solution to this problem must be found, otherwise alternative drilling and sampling techniques need to be developed.

(5) *Conclusion*. The existence of the remotest lake on Earth, near Vostok Station, appears close to being confirmed by geophysical techniques. However, we require a better knowledge regarding the thickness of the water before we can be certain of our conclusions. The program recommended is first to ensure that the existing data is fully exploited (during the next year or two). Planning for further geophysical studies listed in (3) can commence during this period and may be expected to last between 2–5 years. At the same time consideration of the techniques to be used for sampling lake water and sediments, and other means of long-term monitoring, can take place with the aim of carrying out these investigations when associated engineering problems have been solved – possibly starting in 5–10 years.

(Excerpted from the Workshop Report by Robin and Zotikov, 29 November 1993.)

6

A thick water layer exists under Vostok Station!

It was easy to write the word "re-examined" in our workshop report on a revision of seismic data at Vostok, but in reality there was some uncertainty about it as a result of past circumstances. Dr. Kapitsa mentioned publicly that he would do his best in this undertaking, but he also mentioned that he kept the Vostok seismograms in his personal scientific archive at his summer house (which had been partially destroyed in a major fire). We thought, knowing Murphy's Law, that the data would never be forthcoming. However, Lady Luck over-ruled, as she had in many other instances related to Lake Vostok, and Dr. Kapitsa located the Vostok seismograms.

Re-examination of the results of seismic reflections, which were acquired in 1964 at Vostok Station was completed in 1994 in Moscow and Cambridge some 30 years after the experiment. The results showed a very clear P-wave echo from the area of a second reflection, which together with some evidence of a sub-ice lake from radio-echo sounding in this vicinity suggested a water depth of a little more than 500 m beneath the bottom of the ice sheet. This represented a final confirmation of the existence of the lake (Figure 3.2).

In 1994 at the XXIII Scientific Committee on Antarctic Research (SCAR) meeting, held in September in Rome, Dr. Kapitsa presented to various Working Groups of SCAR this new interpretation and the news of a 500 m thick layer of water beneath Vostok Station, thus impressing the delegates and guests at the meeting.

As a result of this discovery, Professor Kotlyakov, Russian delegate to SCAR, wished to draw the Lake phenomenon to the attention of SCAR because of its intrinsic scientific interest and its ramifications for the ongoing ice core drilling program at Vostok Station. He asked other delegates to express their opinions on an investigation of the phenomenon called "Lake Vostok", informing them that this request was connected with the fact that the lake is located just below the Russian station and that for more than 25 years the Soviet (now Russian) expedition had been involved in deep-core drilling there. The depth of the borehole was more than

2,500 m at the time of the meeting, and the total thickness of the ice is 3,750 m. As a result of this discussion and of the Scott Polar Research Institute (SPRI) workshop report of November 1993 the SCAR delegates adopted "SCAR Recommendation XXIII-12" which was addressed to the Russian, French, and U.S. National Committees, the importance of which is evident in the following SCAR recommendation.

SCAR recommendation XXIII-12 concerning the reported subglacial lake beneath Vostok Station

Recognizing the extreme scientific value of the deep ice core drilling at Vostok Station; noting the reported existence of a subglacial lake beneath the drilling site; anticipating that the study of such a lake, possibly isolated from the atmosphere for a million years or more, would also have an extreme scientific value; concerned that such a lake not be inadvertently contaminated by drilling fluids, but at the same time recognizing that the present bottom of the drill hole is still about 1,000 m above the base of the ice sheet.

SCAR recommends to the Russian, French, and U.S. National Committees:

(1) that ice core drilling at Vostok Station proceed to the greatest depth that will still safely ensure undisturbed integrity of the reported lake;
(2) that drilling does not proceed beyond this point until there has been an environmental impact evaluation and relevant scientific investigations have been carried out; and
(3) that a workshop be held soon to consider all aspects of the situation.

The recommendation suggested to continue core drilling in the deep hole at Vostok Station, but mentioned the need for precautions against the penetration of the drilling equipment or a drilling fluid into the lake, or a contamination of the lake by any other means, before a special scientific investigation of this kind of question be completed, technical method of penetration be developed, and the required equipment manufactured.

Following informal consultations, it became apparent that Cambridge, U.K., would be the most suitable location for the proposed workshop. Dr. Heap, the Director of the SPRI at the time, said in his letter of 30 March 1995:

> *Following a suggestion from me, Professor Kotlyakov, Chairman of the Russian National Committee for SCAR, has agreed that the workshop recommended under SCAR XXIII-12 be held in Scott Polar Research Institute from 22–24 May 1995. Professor Miller, Chairman of the SCAR Working Group on Glaciology has agreed to be Chairman of the workshop. The workshop follows the informal meeting in Cambridge on 23 November 1993, which led to the new evaluation of data and to the support given in SCAR XXIII-12. The aim is for a workshop with around 15–20 experts representing a range of interests who can make authoritative statements and recommendations to*

the key National Committees and to SCAR as to the scientific merits of drilling through to the lake and its sediments. Since time is now short I have consulted Igor Zotikov, Gordon Robin, Vladimir Kotlyakov, Claude Lorius, Bob Rutford, and Peter Clarkson (Executive Secretary, SCAR) regarding who should be invited to the meeting ...

7

Deep drilling at Vostok Station

THE STATION

Vostok Station is located directly above Lake Vostok, with access to the lake via a hole drilled from the surface of the ice sheet to 3,663 m depth, leaving a further 130 m of ice remaining before reaching the lake. A review of the station and history of drilling at Vostok Station will aid in explaining the present situation.

Vostok Station is located at a Pole of Cold on our planet. It is called a Pole of Cold because it is the coldest registered place on Earth – with a mean August temperature of −68°C and a minimum temperature in 1974, which is also the world's record low temperature, of −89.4°C. The station is located in the middle of the East Antarctic Ice Sheet. The elevation of the station is about 3,500 m and the thickness of the ice below the station is about 3,750 m. This means that the bedrock elevation is thus below sea level, and the station's elevation is due to the ice thickness only. Rate of precipitation at Vostok is about 2.4 cm yr^{-1}. The area could in fact be called a polar desert because of this low precipitation rate. The temperature of the upper hundred meters increases slightly from −57.13°C at a depth of 48 m. It has a temperature of −53.7°C at a depth of 507 m. The temperature increases slowly at deeper horizons until it reaches the ice melting temperature of −2.4°C for a pressure of about 300 bars at the bottom of the ice sheet. The station has been occupied continuously from 1957 to the present-day (2005), with one exception being the interruption for the polar winter of 1962/1963 when it was closed due to budget constraints. It was re-opened the next summer under pressure from the scientific community, and has been occupied since that time. From 1957 to 1995 tractor trains from Mirny Station provided the main means of food and fuel supply each year. The distance from Mirny to Vostok Station is more than 800 km and it takes more than a month for the train to cover this distance. Russian planes from Mirny Station exchange scientists and station personnel, but cannot bring large amounts of cargo.

From 1996 to around 2000 the Russian Antarctic Expedition (RAE) was unable to supply the station in the traditional way due to financial difficulties associated with the disintegration of the Soviet Union, and the U.S.A. assisted to maintain the station without interruption. A special agreement was signed between Russia and the U.S.A. for a cooperative drilling program and shared use of the ice core for scientific study, with supplies being delivered to the station by American LC-130 Hercules aircraft.

The main purpose of drilling through the ice sheet at Vostok Station up until 1996 was to procure an ice core from different depths to provide a climatic record of Antarctic Ice Sheet temperatures and other data versus time in order to trace climatic changes. The idea was based on two postulates. The first is based on the existence of continuous snow accumulation at the surface of the Antarctic Ice Sheet. Each snow particle does not stay at the surface very long, instead it is buried by other snow particles, which cover it according to the accumulation rate. Snow converts to firn, and firn to ice, and it is possible to determine the time that has passed since the particle was at the surface. This time period depends on the rate of snow accumulation and the depth at which the particle is located. There is a strong correlation between the depth from which the core is taken and the time that has passed since the ice was at the surface of the ice sheet in the form of snow (Nye, 1959; Dansgaard, 1964). The second postulate is based on the correlation between the ratio of stable isotopes Oxygen 16 and Oxygen 18 in the snow at the surface of the ice sheet and the temperature at the surface (Dansgaard, 1964; Dansgaard *et al.*, 1973), as well as the assumption that each particle of ice at any horizon retains the Oxygen 16/Oxygen 18 ratio it had at the surface of the ice sheet.

The first attempt to drill the ice sheet at Vostok Station was made by Station leader Dr. Ignatov in the middle of the 1958 winter (Ignatov, 1960). He attached a weight to electrically heated elements and placed the device on the ice sheet surface. The device penetrated the upper 40 m of the ice sheet, making a borehole, which was dry. Apparently the water escaped from the borehole through permeable walls, but the thermal drill did not penetrate further because meltwater was not removed from the hole at deeper horizons. The heat produced by the drill made the hole larger, increasing its diameter.

It became clear that the meltwater would have to be removed from the hole to maintain downward progress of the thermal drilling device. Under the direction of Dr. Kapitsa, several 40 m deep holes were produced at Vostok Station at the end of 1959 with a rotating device. The author made an attempt in late 1959 to drill a hole at Vostok Station with a thermo-electrical drilling device equipped with a pump to remove meltwater from the bottom of the hole using a special water container in the upper part of the device. However, the pump failed at a depth of about 50 m and drilling was stopped.

DEEP DRILLING OF ICE CORES AT VOSTOK STATION

The first successful attempts to drill deep holes at Vostok Station with thermo-electrical drills equipped with a pump to remove meltwater from the bottom of the hole, also capable of extracting a core, began in 1970.

The Leningrad Institute of Mining and its Department of Drilling, led by Professor Kudryashov, became a main participant in the construction, manufacture, and deep-core drilling at Vostok Station.

The idea of making a borehole by simply melting ice under a hot ring powered by electricity was not only simple but also a natural process. Nobody could imagine then that drilling to the bottom of the ice sheet would take more than 30 years of dedicated work in Leningrad and at Vostok Station itself. Professor Kudryashov, Principal Investigator on the project for 25 years passed away, never having the pleasure of seeing its successful completion (it is interesting to mention that his team, until about 1997, did not believe in the existence of the lake and had prepared themselves and their equipment to reach the bottom and to drill solid bedrock beneath the ice sheet).

The first "dry" borehole drilled at Vostok Station with the thermo-electrical drilling unit in 1970 achieved a depth of 500 m. Temperature measurements showed an increase from −57°C at a depth of 20 m to −53.5°C at 500 m. A total of 293 trips down and up were performed to reach this depth and an average traverse at a single trip (close to a length of the core) was 1.73 m. The average speed of drilling was about 1.3 m h^{-1}. In May 1972 the "dry" borehole at Vostok Station reached a depth of 952.5 m.

The time involved for the drilling tool to reach the bottom at depths of 500 m was significant. It was clear that the borehole was closing under the surrounding pressure and the borehole would have to be filled with some liquid, heavy enough to compensate for the pressure and prevent borehole closure. An aviation fuel including a mix of a type of kerosene, tetrachloric hydrogen, tetrabromethane, and freon was chosen. All components were well mixed, and the resulting mixture had a low temperature stability and did not interact with the water or ice. The viscosity of the mixture was low, which was important because of the hundreds of round trips required to make a deep borehole.

The drilling device consisted of a 10 m long tube of steel with outside/inside diameters of 180/130 mm. The lower end had a ring heating element, which was powered electrically (approximately 3.5 kW, Figure 7.1). The tube was kept vertical on an armored logging electrical cable. When it was placed on the snow or ice surface and the heating device was turned on, the drill began to melt its way downward. A core with a diameter slightly less than the inner diameter of the tube entered the tube until the space was completely filled. Melted water was then pumped from the bottom of the borehole into a section of the tube above the core container. This section served as a meltwater container. The upper part of the tube contained electrical equipment for the drill. When the core container was full, a cable winch at the surface of the ice sheet raised the drill and core. The "teeth" of a specially designed core-cutting device installed at a lower part of the tube cut off the core from the bottom of the borehole and prevented it from falling out while the drill was raised to the top of a drilling mast. The mast was protected from the weather elements by a protective cover. The mast, winch, and logging device with control and monitoring equipment, were mounted inside a drill building, which consisted of two movable steel houses on sleds. The core and meltwater were removed from the

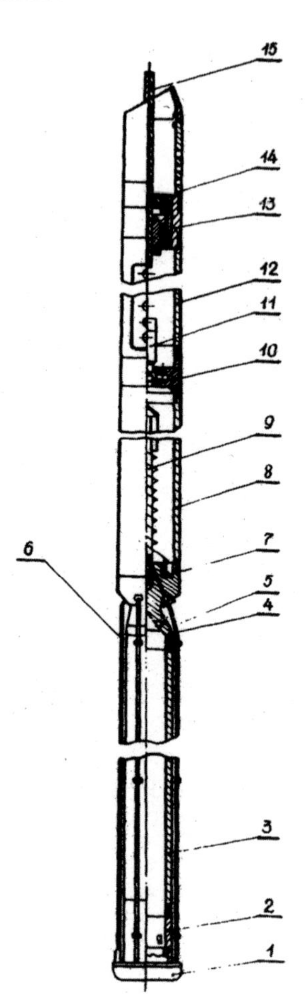

Figure 7.1. The thermo-electric drilling device used to drill deep holes at Vostok Station (adapted from Kudryashov *et al.*, 1982): 1 – heater ring, 2 – core lifter, 3 – core barrel, 4 – drain adapter, 5 – water pipes, 6 – heater power cable, 7 – water heater, 8 – water tank, 9 – water pipe, 10 – pump adapter, 11 – pump, 12 – electrical section, 13 – cable termination, 14 – end cap, and 15 – electro-mechanical cable.

drill and the drill returned to the bottom of the borehole for another round of drilling. The temperature inside the building was maintained above 0°C with the aid of two heaters; the outside temperature was less than −70°C. The temperature inside the core drilling device and the cable just after lifting the device from the hole was below 0°C.

The development, manufacture, and testing of a new drilling device, capable of work in boreholes filled with this liquid, plus the construction of a new 14-m mast and advanced new winch and connecting equipment took more than 5 years. It was not until 1980 that this new drilling installation was transported from Mirny Station

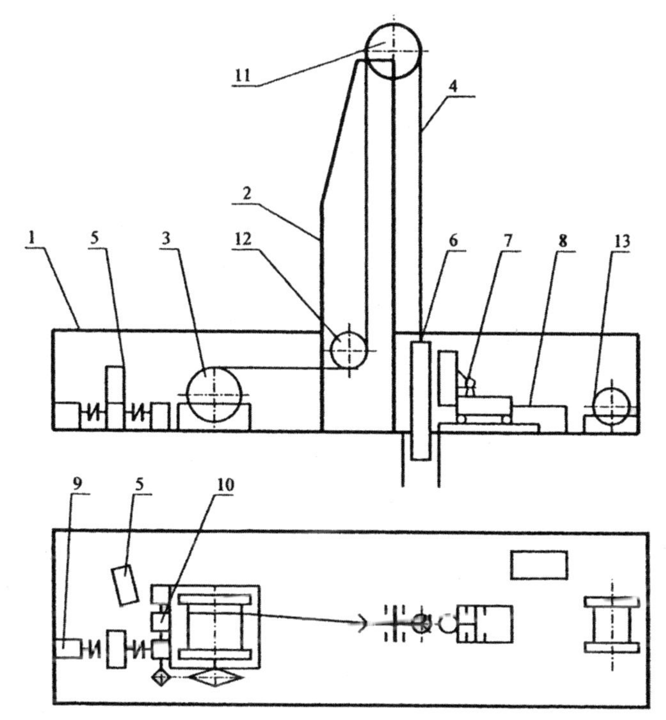

Figure 7.2. Main deep-core drilling complex at Vostok Station (adapted from Kudryashov *et al.*, 2002): 1 – drilling building, 2 – tower, 3 – hoisting winch, 4 – cable, 5 – control deck, 6 – drilling device, 7 – drill-handling equipment, 8 – DC electric power generator, 9 – electric motor, 10 – worm reducer, 11, 12 – pulleys, and 13 – geophysical winch.

to Vostok Station on a tractor train. The filled borehole was drilled using this equipment to a depth of 1,415 m in 1981 and in August 1985, in the middle of the austral polar night, a depth of 2,002 m was achieved. The drilling complex at Vostok Station used for the deepest borehole is shown in Figure 7.2.

The length of the drilling system is 18 m, its width 4 m, and height 15 m. An electrical motor for the winch uses 20 kW, the heating elements 12 kW, and lights 5 kW. The average speed of pulling/lowering operations at a maximum depth of the borehole (up to 4,000 m) with a 16 mm diameter armored cable is $0.7\,m\,s^{-1}$ (Kudryashov *et al.*, 2000).

The drilling of the deep borehole number 5G, with its base 130 m above the lake, was started at the surface in 1990 by the 35th Soviet Antarctic Expedition (SAE) (Kudryashov *et al.*, 2000) using the TELGA thermo-electric drill. A TBZS thermo-electric drill was used for holes filled with liquid for deeper horizons (Kudryashov *et al.*, 2000). In 1993 thermal drilling of the hole was terminated at a depth of 2,755 m.

In 1994 drilling operations were suspended due to financial and logistical difficulties – since that time ice coring and ice core studies have been continued as a collaborative effort between Russia, France, and the U.S.A.

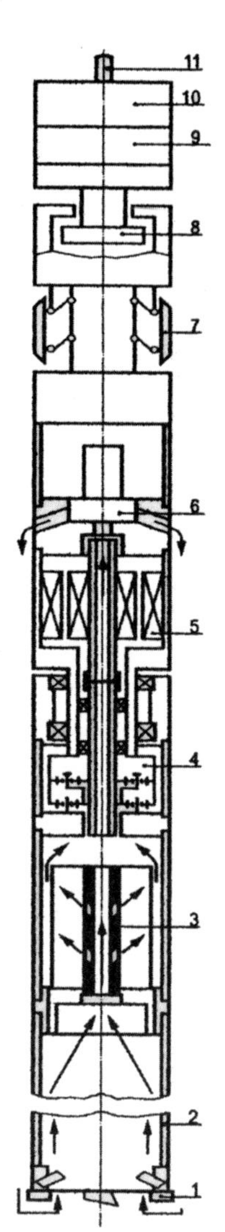

Figure 7.3. Electro-mechanical ice core drilling device KEMS (length 8–13 m, diameter 132 mm, length of core barrel 3 m, weight 240 kg) for drilling the lower part of the deep hole at Vostok Station (from Kudryashov *et al.*, 2000): 1 – drill head, 2 – core barrel, 3 – chip chamber including chip filter, 4 – reducer, 5 – driving electric motor, 6 – pump, 7 – anti-torque system, 8 – hammer block, 9 – electric chamber, 10 – cable suspension clip, and 11 – cable.

The last phase of drilling operations was performed with an electro-mechanical device (KEMS), which was developed and manufactured in Leningrad to drill into the bedrock below the ice sheet under Vostok Station (Figure 7.3). This device had a ring of cutting "teeth" powered by an electrical motor instead of a heated ring and anti-torque leaf springs in the upper part of the drill.

Drilling at Vostok Station was stopped in January 1996, when the depth of the borehole was 3,300 m. Termination of the drilling and conservation of the hole have been undertaken according to a Scientific Committee on Antarctic Research (SCAR) recommendation to prevent any danger of possible contamination of the lake in spite of the fact that some hundreds of meters of undrilled ice remain above the lake water. But under pressure from scientists interested in receiving the deepest possible ice core for study, sessions of core drilling proceeded in the 1997–1998 field season (43rd Russian Antarctic Expedition), and hole 5G reached a depth of 3,623 m, which remained the situation in autumn 2004.

It was shown before the termination of drilling that stable isotopes, dust, and electrical conductivity measurements (ECM) are well preserved in the ice from the surface of the ice sheet down to the depth in borehole 5G, offering a continuous climate record for the last 400,000 years. It was shown that there were four distinct climatic periods of cooling and warming within this period of time. Results of the study of this core by specialists in Russia from the St. Petersburg Arctic and Antarctic Research Institute and Mining Institute and the Institute of Geography of the Russian Academy of Sciences, and in France at the Laboratoire de Glaciologie et de Géophysique de l'Environnement, Grenoble and the Laboratoire de Modélisation du Climat et de l'Environnement, Saclay under the framework of Russian–French collaboration on Vostok core investigation are remarkable. Some results are shown in Figures 7.4 and 7.5. It was also shown that close to this depth the isotopes and ECM signals became smooth and were no longer an effective means for deciphering the glacial–interglacial changes. This meant that a two-decade-long core which provided access to the most extended and undisturbed paleo-environmental series extending 400,000 years, along with the last four climatic cycles, came to an end.

ACCRETED ICE UNDER VOSTOK STATION

It was mentioned above that on agreement with SCAR, a new round of drilling, with the purpose of collecting an ice core, was started as a collaboration program between Russia, France, and the U.S.A. This stopped at a depth of 3,623 m, about 130 m above the lake. New and interesting information developed as a result. A very sharp and significant variation in stable isotope and dust concentrations around the 3,320 m depth were found, which could not be of climatic origin (Figure 7.6). The ice stratum from 3,538–3607 m displays visible inclusions of millimeter size, which may originate from bedrock (it is difficult to date this depth). However, in general this layer surprised investigators – it had very large ice crystals, meters in size (the size was measured in the vertical direction, along the length of the core). Electrical conductivity of this layer was two orders of magnitude lower than in previous layers, but the stable isotope content was only slightly different. It was clear that a new area in the ice had been reached, but what did it represent?

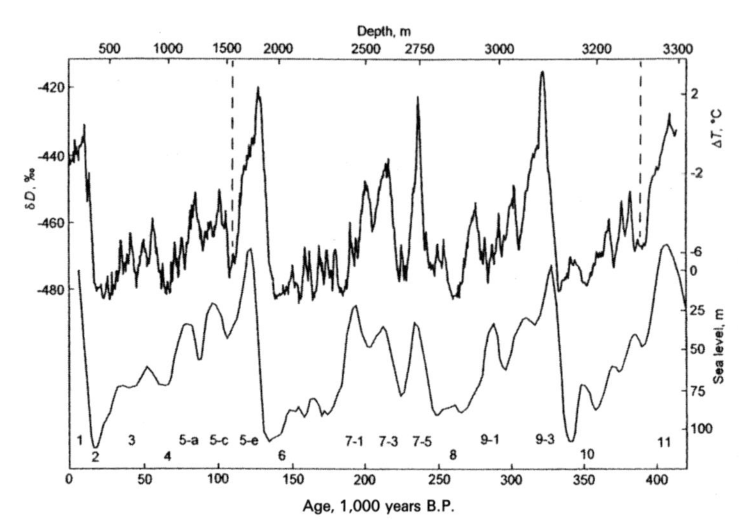

Figure 7.4. A change in temperature of the Antarctic Ice Sheet surface over the course of 420,000 years (adapted from Kotlyakov and Lorius, 2000). Upper curve was reconstructed on the basis of "$\delta^{18}O$‰" changes (deuterium profile) measured in ice cores from deep borehole 5G at Vostok Station, taken from different depths (upper scale) or ages of cores (lower scale). Lower curve shows variations of the total ice volume on the Earth, reconstructed on the basis of the profile of "$\delta^{18}O$‰" interpreted from ice cores from the 5G borehole core. Four total 100,000 year long climatic circles of the Earth's history are clearly visible.

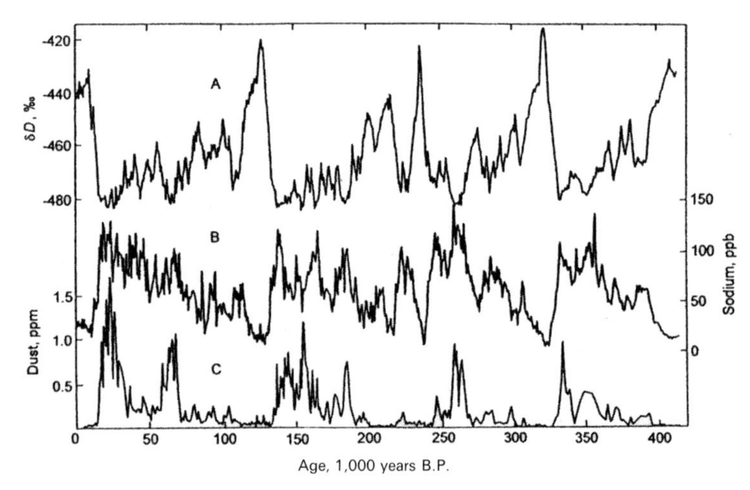

Figure 7.5. Change in temperature and content of particulates in ice cores noted as a result of the study of the 3,330 m deep core from borehole 5G at Vostok Station (adapted from Kotlyakov and Lorius, 2000). (A) Deuterium profile. (B) Sodium profile obtained from samples every 3–4 m on average. (C) Dust profile for every 4 m on average – the dust concentration is expressed in parts per million on the assumption that Antarctic dust has a density of 2.5×10^3 kg m^{-3}.

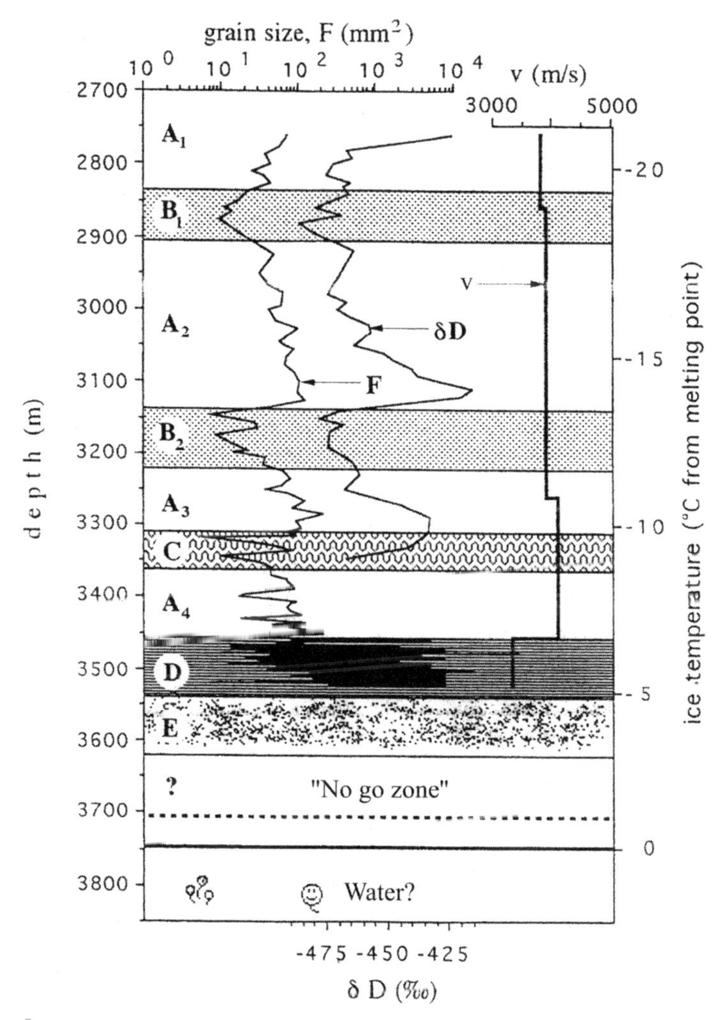

Figure 7.6. Internal structure of the bottom part of the Antarctic Ice Sheet at Vostok Station as revealed by deep-core drilling (adapted from Lipenkov and Barkov, 1998; Souchez *et al.*, 2000). Ice above the 3,310 m horizon is glacier ice with a meteoritic signature. Ice below 3,310 m (down to 3,539 m) is deformed glacier ice. Ice below the 3,539 m horizon to the 3,609 m horizon is accreted frazil ice with visible dirt inclusions. Ice below 3,609 m down to the bottom end of the core (3,623 m) is accreted frazil ice without visible dirt inclusions (Lipenkov and Barkov, 1998; Souchez *et al.*, 2000).

It was a surprise when from the isotope studies, ECM, and gas content measurements it became clear (Jouzel *et al.*, 1999) that the lowest (below the 3,538–3,539-m horizon) part of the ice core was composed of accreted ice (i.e., ice possibly formed from the frozen water of Lake Vostok). This part of the layer is highlighted using the letter "E" in Figure 7.6.

SEARCH FOR LIFE IN THE ICE BELOW VOSTOK STATION

Since 1976 scientists at Vostok Station have been working on solving the question of whether living material exists in Lake Vostok. The first ice cores for biological (actually microbiological) analyses were taken at Vostok Station by Dr. Abizov from the Institute of Microbiology of the Russian Academy of Sciences, who retrieved the first ice cores. A portable Microbiological Drilling Unit and a laboratory for sterile sampling and analyses of ice cores was constructed. A small structure (9 m × 3.7 m and 2.5 m high) was installed on a sled and situated 400 m away from the station to the west (in the direction opposite to the main wind direction). The microbiological house was equipped with an electro-thermal (later mechanical) drill. Some procedures were taken to sterilize the drill and its components prior to the beginning of drilling (Figure 7.7). The core drilling for microbiology was only conducted from within this structure. The samples from the core were extracted and their inner parts were removed using a special device constructed for this purpose (Figure 7.8).

Micro-organisms were found and samples from the core were extracted to a depth of 2,570 m. Thawed samples were placed in sterile flasks that were immediately sealed and refrigerated. The flasks were delivered to the Institute's laboratory, opened in a sterile chamber, and their contents inoculated into various nutrient media. Parts of the samples were passed through a filter with a pore diameter of 0.2 μm to precipitate microflora, other organisms, and various particles whose size exceeded the filter pore diameter. An epifluorescence method and SEM were also used. Micro-organisms of several kinds were found at several horizons, including those belonging to different taxonomic groups characterized by considerable

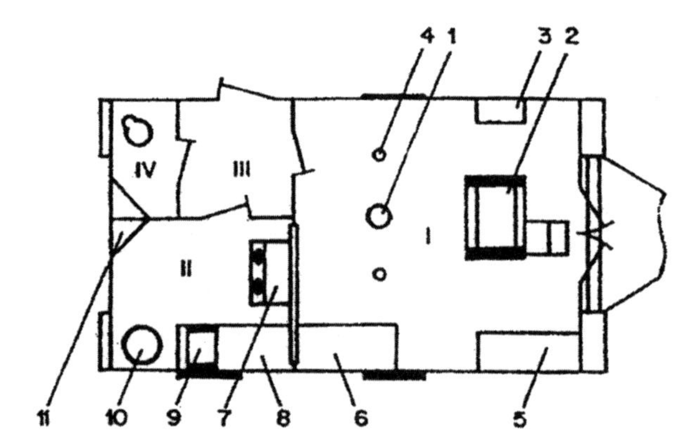

Figure 7.7. The housing for the microbiological deep-core drilling at Vostok Station (adapted from Abizov, 2001): I – drilling section, II – microbiological section, III – entry, IV – washing space, 1 – the mouth of the borehole, 2 – the winch, 3 – drilling control, 4 – foundations of the drilling rig, 5 – table, 6 – storage box for sterile extraction of ice core samples, 7 – box for sterile ice sample analyses, 8 – working table, 9 – thermostat, 10 – hot steam pot, and 11 – water tank.

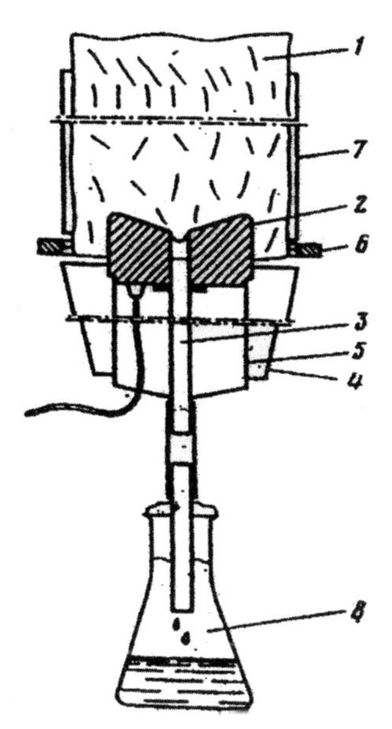

Figure 7.8. A device for sterile extraction of ice samples from the ice core station (adapted from Abizov, 2001): 1 – ice core, 2 – heater, 3 – sterile meltwater pipe, 4 – water collection cup, 5 – support for heating element, 6 – ring of breaking device, 7 – ice core support, and 8 – sterile bottle.

morphological diversity. Prokaryotic micro-organisms (bacteria of various shapes and sizes) were found throughout the whole length of the core. Figure 7.9 shows these features in ancient ice cores located at depths from 1,500 to 2,500 m (Abizov *et al.*, 1998b).

Dr. Abizov showed by using the results of inoculating the samples into nutrient media, that the microflora of these horizons chiefly consisted of spore forming bacteria (Abizov, 1993). It is possible that the vegetative cells of spore forming bacteria, which represent the majority of grain-positive bacteria, are especially cold resistant and capable of long-term anabiosis with their vital functions being retained (Abizov, 1993). The microbial cell numbers in different horizons ranged between 0.8×10^{-3} and 10.8×10^{-3} in 1 m^3 of ice.

Figure 7.10 (Abizov, 1998b) shows the quantitative distribution of micro-organisms and dust particles in ice samples from different depths, corresponding to different ages of ice. It seems that there is no definite law underlying the quantitative distribution of these organisms with respect to depth. However, glaciological studies of dust particle distribution along the ice core length show that the fluctuation of dust concentration depends on the fluctuation of climatic conditions (Jouzel *et al.*, 1993).

Dr. Abizov concluded from the data that microbial cells occurred at a depth of 1,500–2,759 m in the ice sheet below Vostok Station, and have remained viable for 110,000 – 240,000 years, establishing for the first time that micro-organisms could

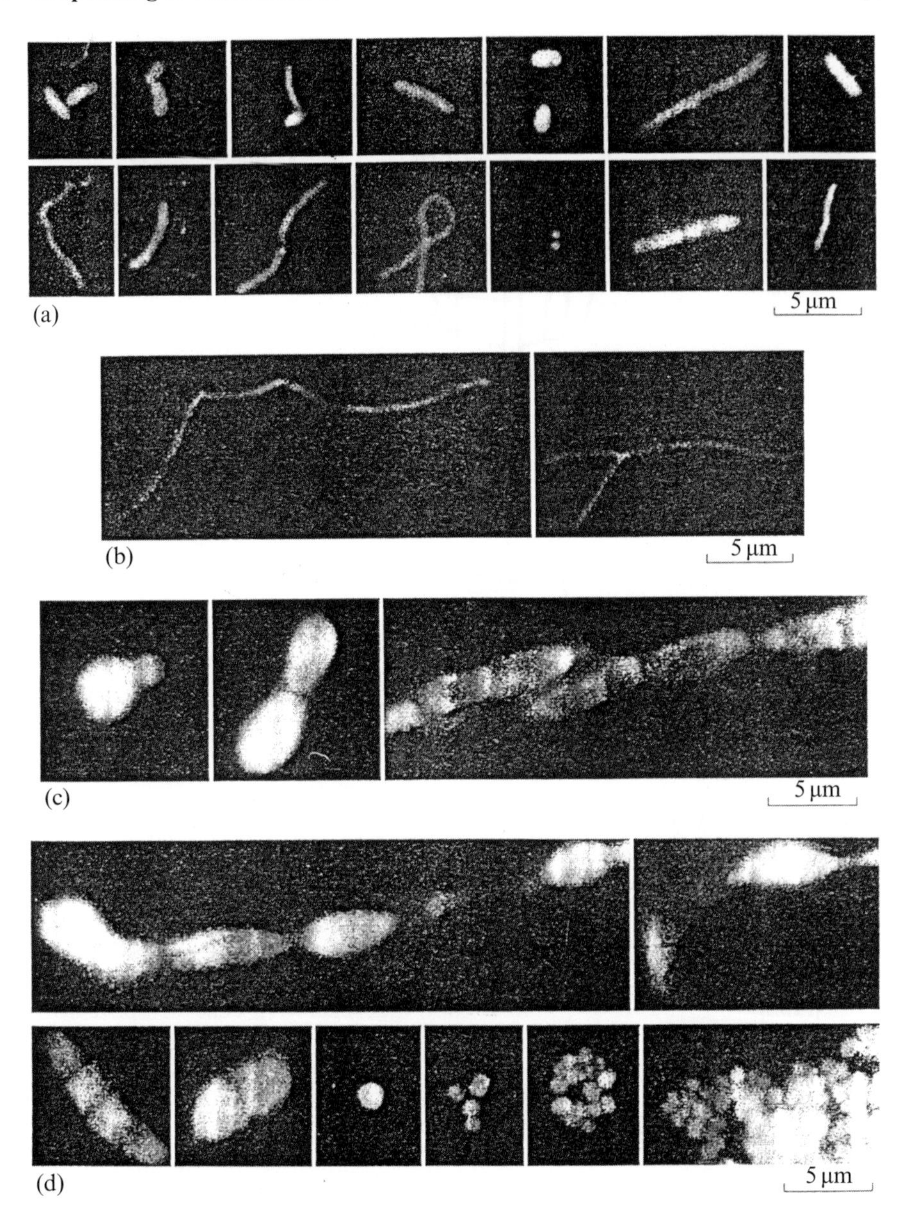

Figure 7.9. Microflora of the ancient ice horizons below Vostok Station: (a–e) and (g) – fluorescent and (i) electron microscopy. (a) Bacterial cells, (b) fragments of actinomycete cells, (c) budding and dividing yeast, (d) partially lysed hyphae of fungi (upper row) and conidia of various fungi (lower row); (e) and (f) microalgae, (g) dust particle adsorbed bacteria (adapted from Abizov *et al.*, 1998b).

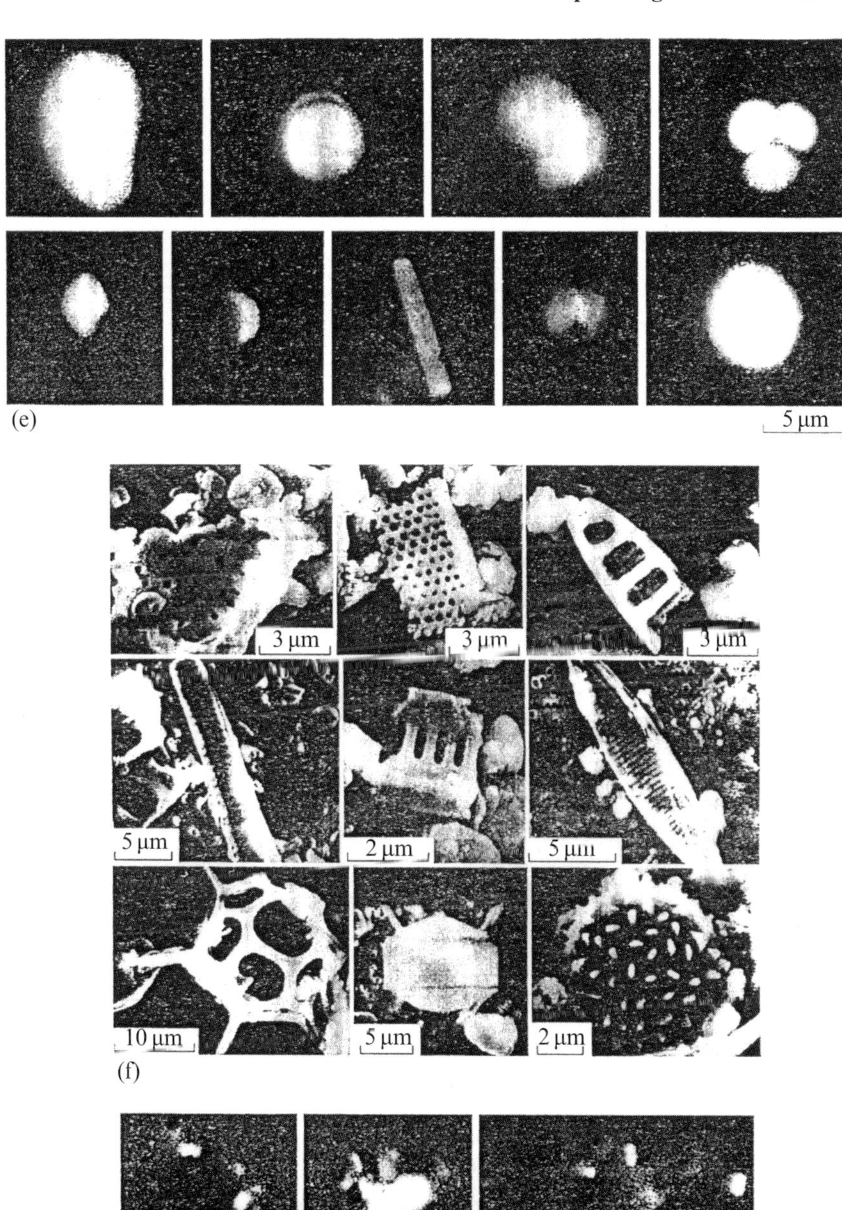

(e)
(f)
(g)

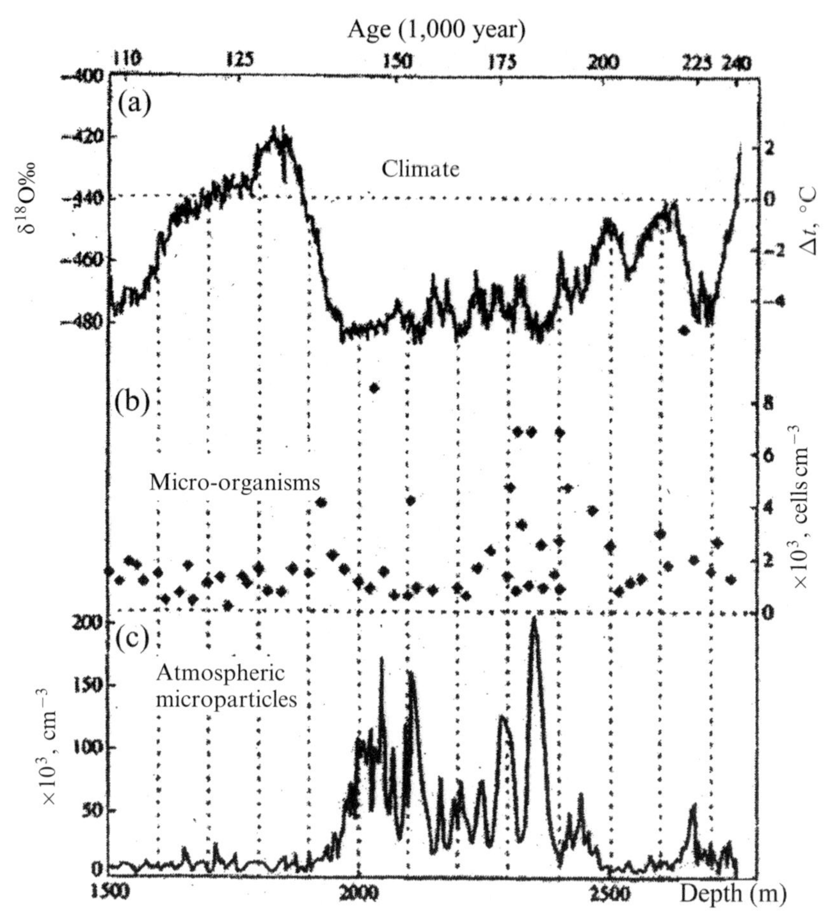

Figure 7.10. Distribution of numbers of micro-organisms and dust particles along the length (age) of the Vostok Station ice core (adapted from Abizov *et al.*, 1998b). (a) Climate change, (b) concentration of microbial cells along the core, and (c) concentration of dust particles along the core.

exist in anabiosis for this period of time in Antarctic ice with their vital functions retained. However, the number of viable cells decreases with increasing depth (and age) and does not correlate with the total number of intact cells at the same horizons (Figure 7.11).

The discovery of accreted ice at the bottom of the ice sheet opened the way for a microbiological study of Lake Vostok water through ice core studies (Priscu *et al.*, 1999; Karl *et al.*, 1999; Abizov *et al.*, 2001; Bulat *et al.*, 2001) because there are 130 m of the ice sheet remaining between the bottom of the borehole and the ice/lake water interface. However, the danger of contamination of the lake is a major issue preventing further drilling.

There is evidence supporting the existence of two distinct layers in the accreted ice stratum. The upper 70-m layer (3,538–3,609 m) containing visible mineral

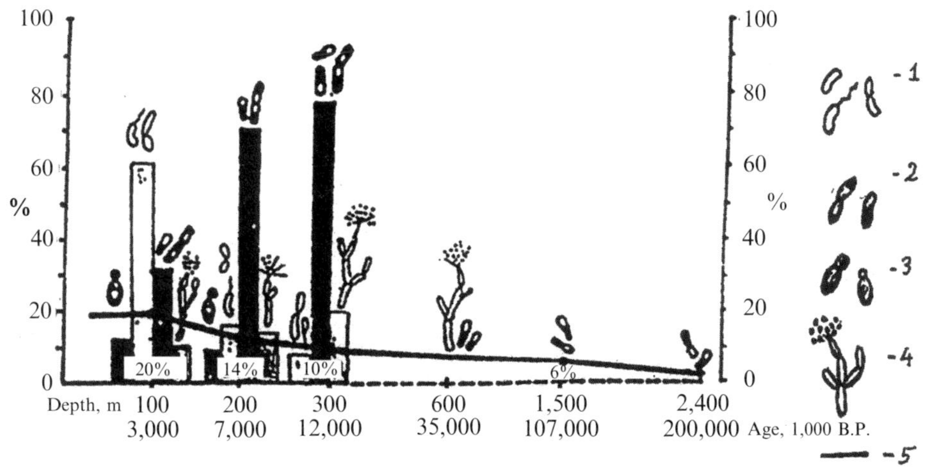

Figure 7.11. Percentage of vital micro-organisms of different types decreasing with depth (and age) of the ice core (adapted from Abizov, 2001): 1 – non-spore forming bacteria, 2 – spore forming bacteria, 3 – yeasts, 4 – mycelium mushrooms, and 5 – a change of the total number of vital micro-organisms.

inclusions is thought to have been accreted over turbulent lake water (Wuest and Carmack, 2000). The isotope and gas content data (Souchez *et al.*, 2000; Lipenkov and Istomin, 2001) suggest that formation of this upper section of the lake ice was accompanied by trapping and subsequent freezing of liquid water inclusions (so-called "water pockets"). On the other hand, the extremely low and uniform gas content of the ice, and the absence of mineral inclusions in the lower section of the accreted ice stratum (from 3,609–3,623 m and likely down to the ice–water interface) rule out significant involvement of liquid water pocket formation in the freezing process. This deepest ~140-m section of the Vostok ice has likely been accreted at the maximum water depth where the convection of lake water was very low (Wuest and Carmack, 2000). Consistent with this, both the gas content and the concentrations of major ions are found to be a few orders of magnitude higher in the upper layer of the accreted ice than in the lower layer.

It was mentioned previously that the preliminary study of this accreted ice from different horizons shows the existence of different micro-organisms in the core. The number and diversity of these micro-organisms is different for each of the different horizons and correlates with organic and inorganic inclusions in the ice samples, but there is no distinct evidence that the number and types of organisms in accreted ice differ dramatically from ice of glacial origin.

The study of accreted ice included studies of the chemistry of the lake water and presence and types of organisms that might be expected in Lake Vostok. Extrapolation of data from the accreted ice study may be questioned (e.g., by the large ice crystal structure and unusual grain growth that creates low porosity, which restricts the infusion of liquids into the crystal lattice). This means that fluids at grain boundaries may not be representative of chemical impurities in the large mass

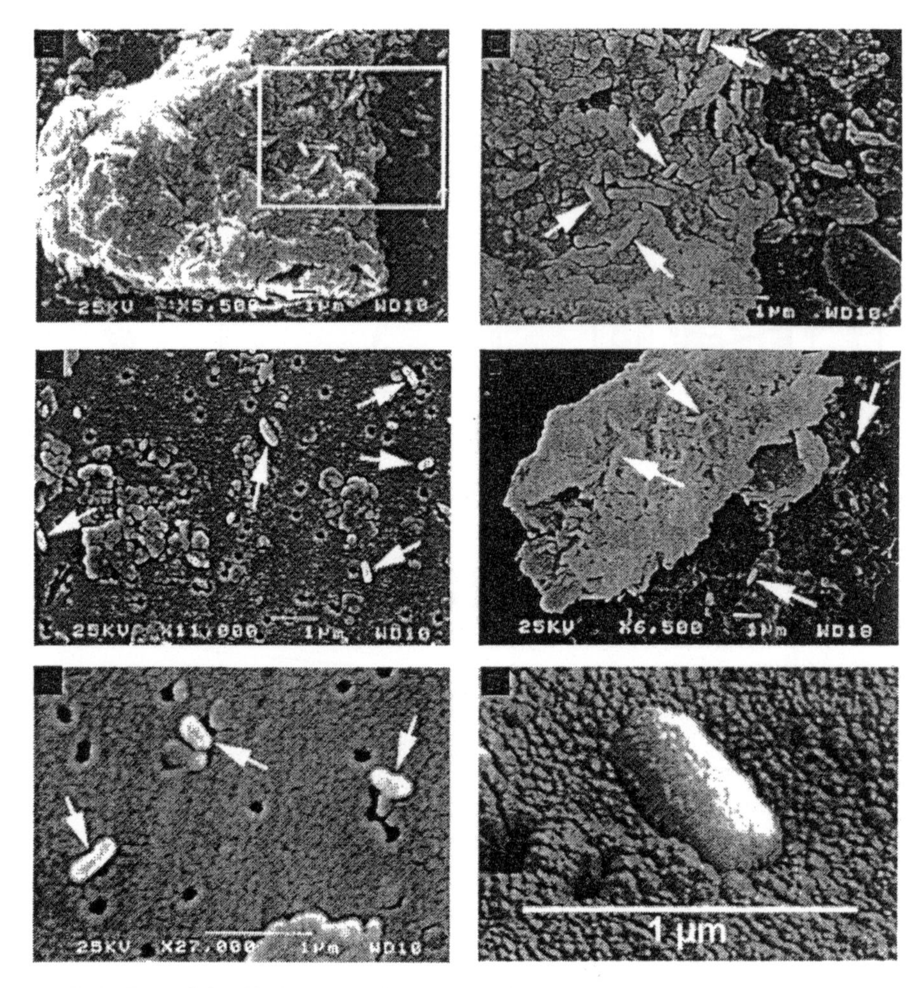

Figure 7.12. Bacterial cells in accreted ice taken from a lower part of the Vostok Station borehole (adapted from Priscu *et al.*, 1999). Cells (marked by arrows) under scanning electron microscope (A–E) and atomic force (F) magnification.

of ice (Montagnat *et al.*, 2001). As a result, the use of accreted ice composition to predict lake water composition has to be viewed with caution. However, models predict that active microbial assemblages should be able to exist in veins and interstices between ice grains.

A number of microbes have been detected in accreted ice, and data reveal that bacterial diversity is low with the DNA detected being typical of modern DNA (Priscu *et al.*, 1999; Karl *et al.*, 1999). Figure 7.12 shows (arrows) bacterial cells identified by Priscu *et al.* (1999) using a scanning electron microscope (Figure 7.12 (A–E)) and an atomic force microscope (Figure 7.12(F)).

However, molecular biological studies of this ice (Bulat *et al.*, 2001) suggest that there are few if any microbes in the accreted ice, and most microbes detected until now may be related to contamination of the drilling fluid and core handling. Some new data acquired by "ultraclean" technology shows that bacterial densities in previous reports were overestimated by perhaps an order of magnitude. It was also shown that a number of viruses exist in glacial and accretion ice and these viruses show a variety of unusual morphologies. The origin of these viral particles is unknown, but DNA studies to identify it are underway. Siegert *et al.* (2003) believe that microbes associated with accreted ice migrated through the ice and entered the surface waters of Lake Vostok in meltwater being quickly incorporated in the accretion process before the meltwater mixed with deeper lake water. Some molecular biology studies identified a strain of bacteria whose DNA is similar to hemophilic bacteria (Bulat *et al.*, 2001), suggesting the possibility of hydrothermal activity in Lake Vostok. If proved, it will provide new ideas on what to expect in the waters of Lake Vostok (Bulat *et al.*, 2004).

8

Vostok Station and the ice shelf

A second workshop on the study of Lake Vostok occurred in May 1995 at the Scott Polar Research Institute (SPRI), University of Cambridge, England, upon the recommendation of SCAR. A grant from the Royal Society in London provided me with support whilst at the SPRI in order to prepare for the meeting some 6 months in advance. Dr. Robin, former director of the SPRI, provided access to archives of all the radio-echo sounding flights that he conducted in East Antarctica. A reel from the archives was examined that had more than 100 negatives taken from an oscilloscope screen that showed a return radio-echo with coordinates of signal intensity versus time when the signal returned – so-called "A-scope" pictures. This reel was of special importance because the pictures were taken on 1 January 1974 on flight Number 130, the only flight along the entire length of Lake Vostok (Figure 8.1).

According to Dr. Robin the main equipment for continuous monitoring of the ice sheet bottom did not work properly during this flight, as it was jammed for most of the time the plane flew over the lake. Only a small part of the flight shown in this picture was marked with thick, solid lines representing the ice–water interface marked "lake" at the left-hand side of Figure 4.2, and dashed lines representing ice/dry bedrock interface at the right side of Figure 4.2. However, equipment that produced "A-scope" pictures worked properly for the entire time the plane flew over the lake. Black dots in Figure 8.1 correspond to places where the "A-scope" pictures show the ice–water interface of the kind marked "lake" at the left-hand side of Figure 4.2 (see Figure 8.2).

A large, sharp signal at the left-hand side of each "A-scope" picture can be seen in Figure 8.2, interpreted as a reflection from the air–upper ice sheet surface interface. Farther to the right are reflections from internal layers, with intensities decreasing steadily to the right. A sudden increase of intensity of reflections at the far right represents the bottom of the ice sheet. These bottom reflections are narrow and high, with sharp peaks. Dr. Robin interpreted these reflections to be the ice–water

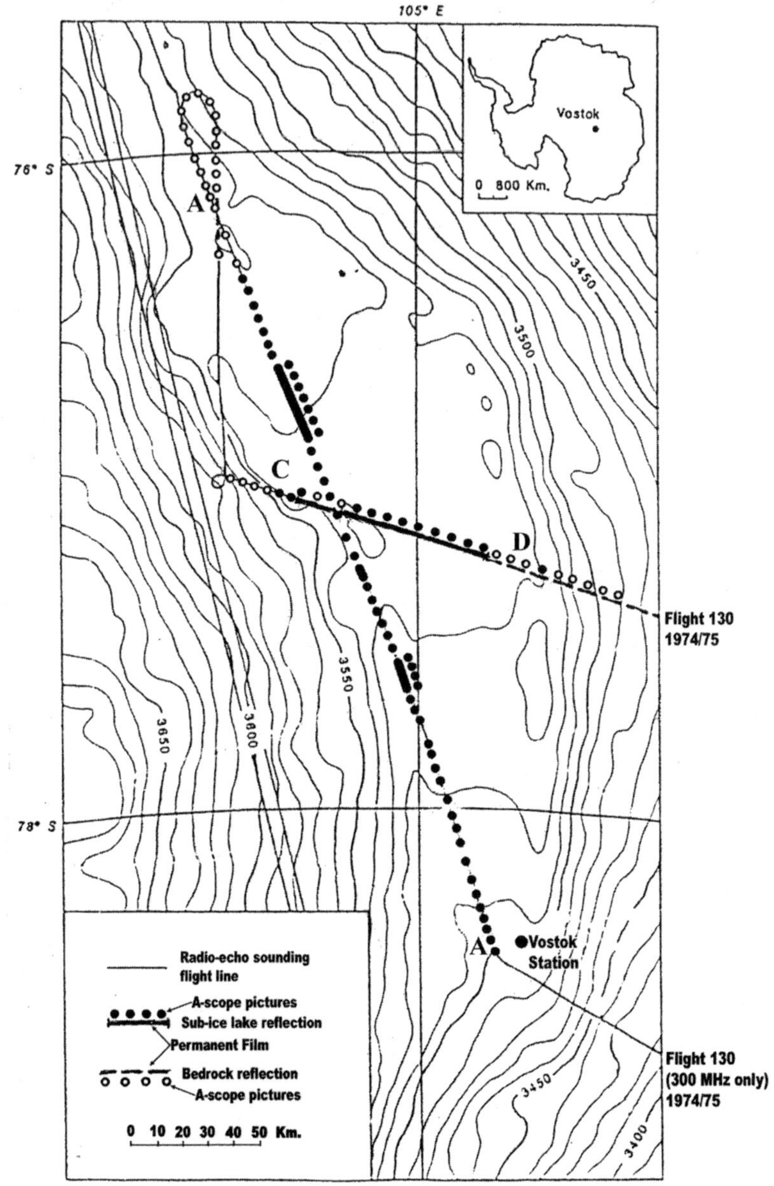

Figure 8.1. Lake Vostok seen as the relatively flat area of ice surface elevation contours. The thin, long straight line represents the route of radio-echo sounding for flight number 130 (1 January 1974), the only radio-echo sounding flight along the long axis of Lake Vostok. Thick, solid parts of this line show where the ice–water interface was recorded on a film of permanent bottom reflections of the type seen in Figure 4.2. Dashed parts of the line show where the film indicates the ice–bedrock interface. Dark dots correspond to separate "A-scope" pictures showing the ice–water interface. Open circles correspond to separate "A-scope" pictures showing the ice–bedrock interface.

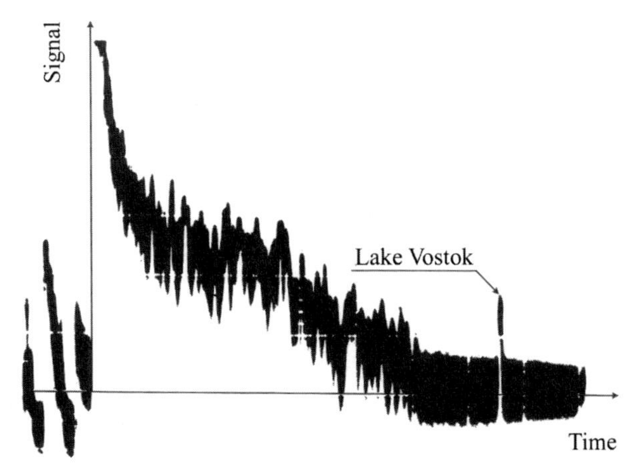

Figure 8.2. An "A-scope" picture. A radio-echo sounding picture from an oscilloscope screen, taken above Lake Vostok, showing a return radio-echo signal from the ice–water interface with coordinates of intensity of signal versus time when this signal returned (Dr. Robin, private commun.). It should be mentioned that the sharp, slim signal marked as the "Vostok Lake", travels to the surface from Lake Vostok in about 47 μs.

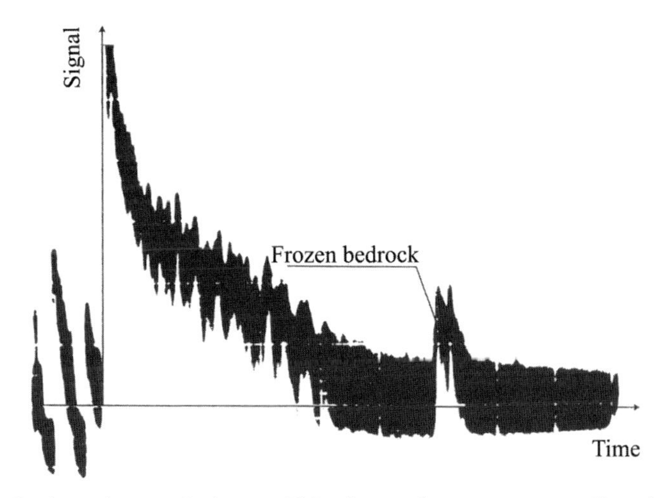

Figure 8.3. A further "A-scope" picture. This picture shows a return radio-echo signal from the ice–bedrock interface. Note the weak, wide signal, which comes to the surface from areas where the ice sheet is on "dry" bedrock (Dr. Robin, private commun.). Travel times to the surface are approximately 30 μs.

interface, thus providing further evidence for the presence of Lake Vostok beneath the ice (black dots in Figure 8.1).

The left-hand side of the "A-scope" picture in Figure 8.3 is similar to what is seen in Figure 8.2, but the right-hand side is different. There is a sudden increase of intensity of reflections as in Figure 8.2, but the increase is not as high nor as sharp as the reflection from an ice–water interface.

Dr. Robin interpreted the latter as a signal from the boundary of the ice sheet and the underlying bedrock. Locations with this type of bottom reflection along the track of flight number 130 are marked by open circles in Figure 8.1. Note also that Figure 8.1 shows that a large part of flight number 130 is marked with black dots, indicating that the plane flew along the long axis of Lake Vostok as well as across it. In only one place, the middle of the lake, did the reflection of the ice–water interface change its appearance, becoming shorter and thicker, suggesting that the plane crossed an island or a peninsula (further study confirmed this was the case).

In later years (Popkov *et al.*, 1998; Masolov *et al.*, 2001), the combined study of seismic and radio-echo sounding along line 2–2′ (Figure 9.1) east of Vostok Station showed that the "A-scope" reflections for ice–water and ice–bedrock interfaces confirmed Dr. Robin's interpretations of Figures 8.2 and 8.3 (see Figure 8.4).

The common feature of all "A-scope" pictures is the monotonous change of the distance between the upper and the lower signals, which corresponds to the thickness of the ice above the lake (a line of black square dots in Figure 8.5). This ice thickness increased from 3,700 m at the southern edge of the lake near Vostok Station, where the plane approached the lake, to about 4,200 m near the northern shore, where the plane left it as seen in Figure 8.5, representing a difference of about 500 m.

It can be seen from the map of surface contours of the ice sheet above the lake in the lower part of Figure 8.5, that the surface elevation of the ice sheet near Vostok Station is 3,500 m and increases slightly to the southern edge of the lake, a difference of less than 50 m. It is clear that the floating ice cover moves, spreading from north to south, because only by such movement and spreading of the ice can it support nearly 500 m difference of floating ice cover thickness at the northern and southern parts of the lake. The similarity between floating ice cover exhibiting features mentioned above and ice shelves in glaciology is apparent. Hence it is reasonable to give the name "Lake Vostok Ice Sheet" to the part of the ice sheet above the lake.

There are some features common to all ice shelves. The first common feature is the water mass circulation in the subglacial caverns under the Lake Vostok Ice Shelf. The circulation is driven by the difference in elevation of the ice–water boundary along the ice shelf – when a change in water temperature along this boundary is determined by a change in freezing point due to a change in water pressure along the boundary. This kind of circulation is similar to that found beneath the Ross Ice Shelf. In the case of Lake Vostok it is expected that the water rises on one side of the lake and moves down on the other, with horizontal currents of opposite direction existing at the lake bottom and roof. This process can influence the intensity of biological life in different places of the lake and has to be taken into account when choosing favorable places to search for life. Vertical (both upward and downward) movement of lake water in different places would redistribute the geothermal heat flow, which rises from within the Earth and eventually reaches the bottom of the ice sheet, thus producing changes in the heat balance at the bottom. In such a situation, it is possible to expect local areas of increased bottom melting in some places and decreased melting or localized freezing of the lake water at other places. This ice shelf differs from those more commonly found at

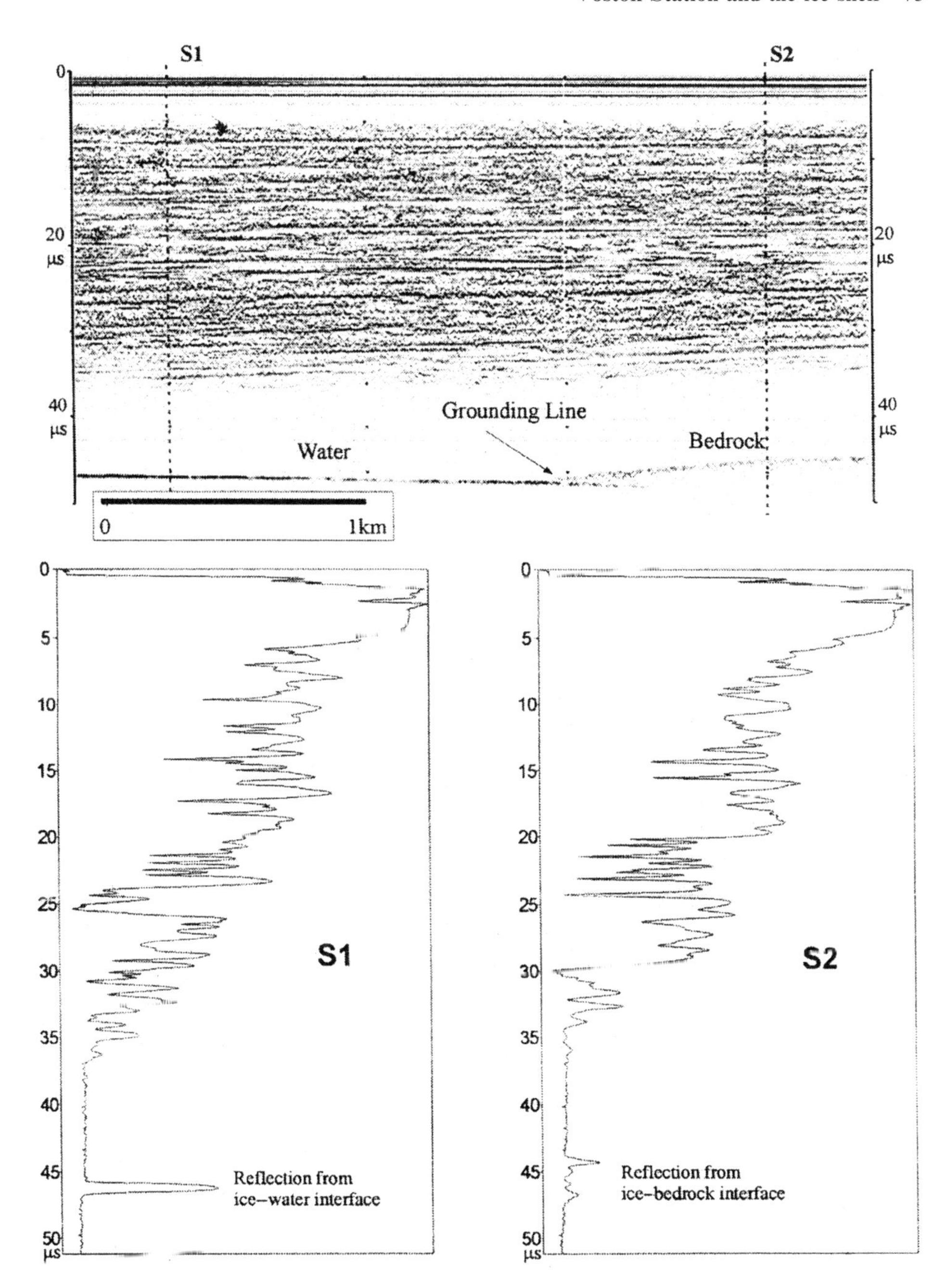

Figure 8.4. Example of radio-echo sounding through the ice sheet above Lake Vostok (line 2–2′ east of Vostok Station as shown in Figures 9.1) and selected radio-echo sounding traces ("A-scope"-type pictures) for ice–water and ice–bedrock boundaries confirming Dr. Robin's interpretations of Figures 8.2 and 8.3 (adapted from Masolov *et al.*, 1999).

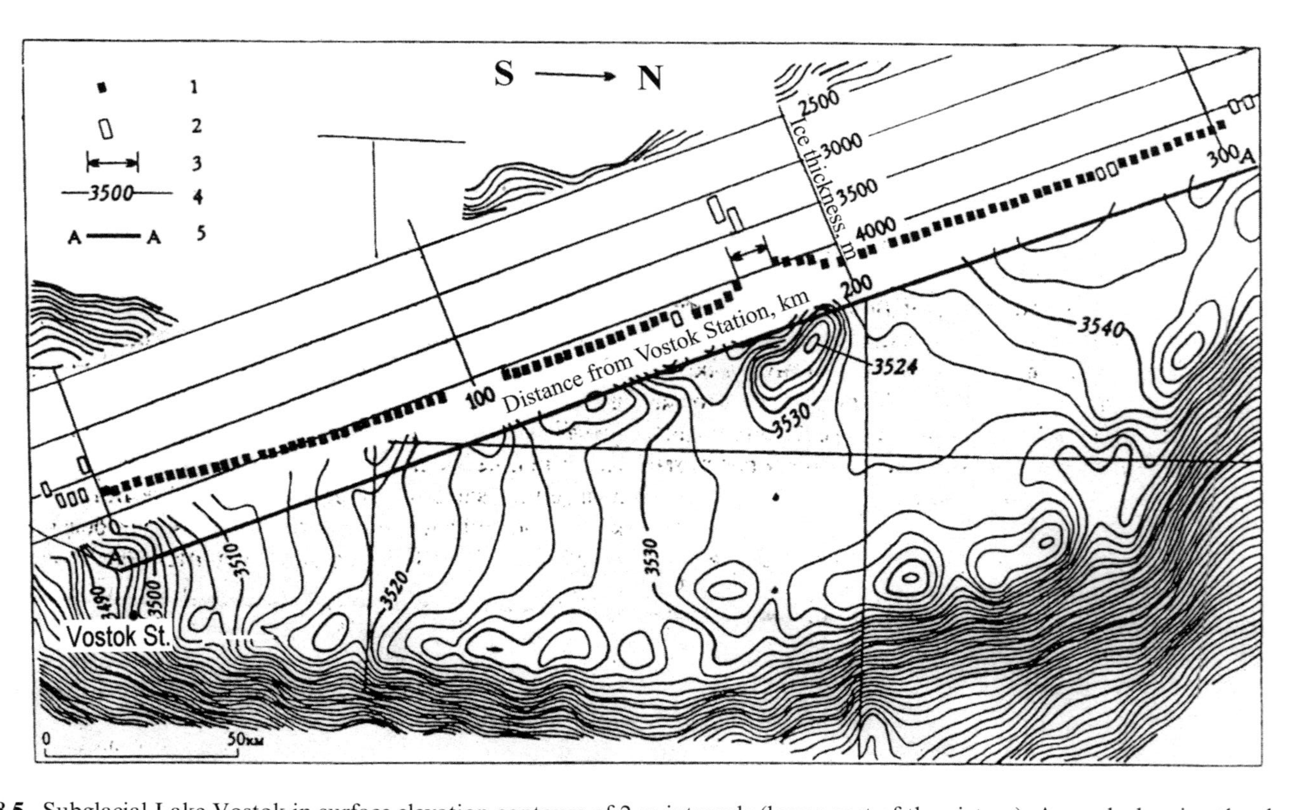

Figure 8.5. Subglacial Lake Vostok in surface elevation contours of 2-m intervals (lower part of the picture). A graph showing the change of Lake Vostok ice cover thickness along the length of the Lake, obtained from "A-scope" picture studies is shown in the upper part of the picture (adapted from Zotikov, 1998). (1) Black square dots represent positions of "A-scope" ice–water interfaces along the Lake; (2) open square dots represent positions of "A-scope" ice–bedrock interface (heights of the squares indicates the uncertainty of the interface positions from the "A-scope" picture studies due to the width of interface reflections); (3) island or peninsula in the middle of the lake; (4) surface elevations (meters above sea level); and (5) – line of flight number 130 (the graph is plotted along this line).

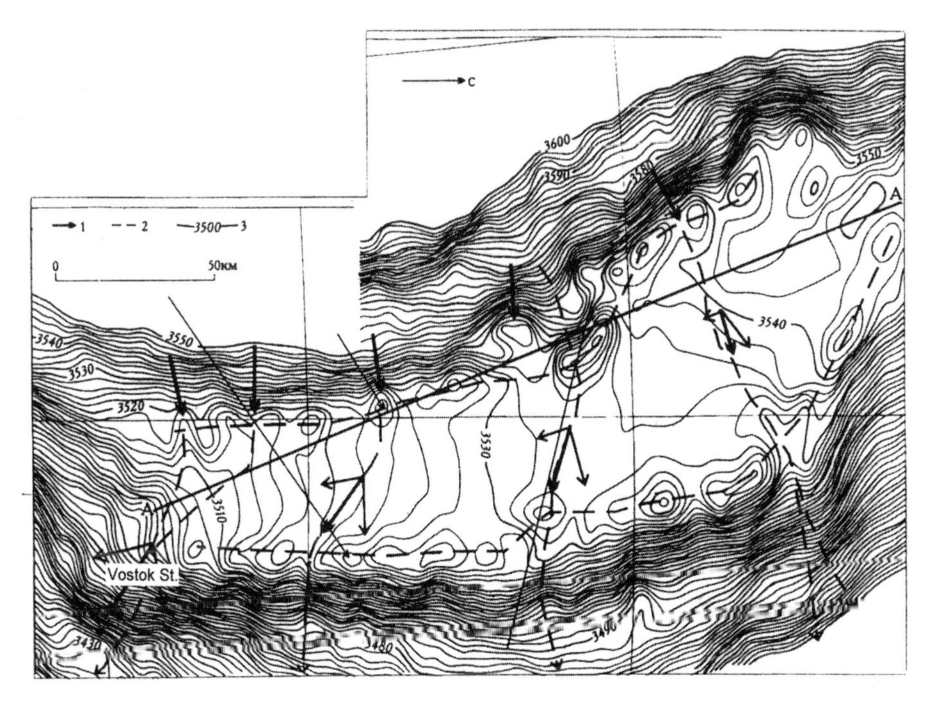

Figure 8.6. Lake Vostok Ice Shelf in surface elevation contours of 2 m. (1) Vectors of ice cover movement; (2) Lake Vostok Ice Shelf shores; (3) surface elevations (meters above sea level). Note the "along lake" component of movement of this ice shelf. This component increases from north to south.

the perimeter of the coasts of Antarctica. First, it is much larger than modern ice shelves, although there is evidence that large ice shelves existed in the past. Second, this ice shelf does not have an open side that marks an open water level. This ice shelf should thus be called an "internal" ice shelf and an "open" water level should be determined by calculation.

If the ice cover above the lake is an ice shelf, then the component of movement of this ice as an ice shelf is directed from north to south and should increase in this direction (Figure 8.6).

The differences of ice thicknesses and surface elevations in different parts of the lake's ice cover were studied to estimate the water density and salinity of Lake Vostok without actual penetration into the lake. Investigations by the author have shown that for an ice shelf, where each part of its body is in hydrostatic equilibrium with the water below it, there is a simple theoretical relation between the change of the surface elevation H and the thickness h of the ice shelf and the ratio of the mean density of ice shelf ("ρ_{ice}") and density of the water, in which the ice shelf floats ("ρ_{water}"). This relationship can be represented by plotting H on the vertical axis for different locations on the ice shelf against h on the horizontal axis – theoretically this should plot a straight line. In this case the slope of the line represents a ratio of the mean ice shelf density to water density (Kapitsa *et al.*, 1996).

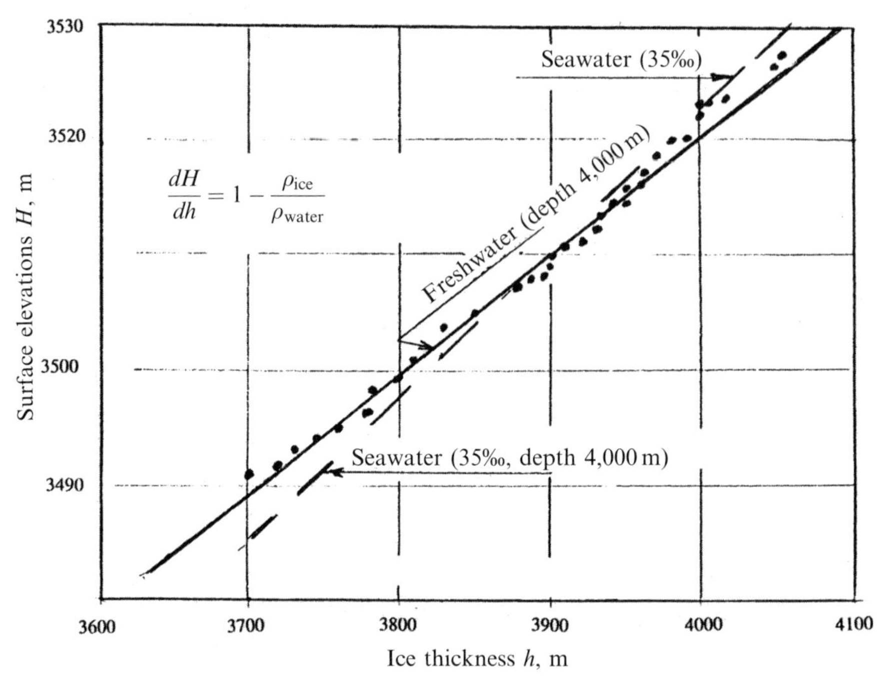

Figure 8.7. Thickness h of floating ice from "A-scope" pictures plotted against surface elevation H at locations clear of the boundary effects along and across the lake for flight number 130. Also shown are theoretical slopes for ice in the same thickness range given by the equation in the figure, where ρ_{ice} and ρ_{water} are densities of the ice and the water of the lake, respectively. One of these lines represents water of 35‰ salinity (pure seawater), the other for freshwater (adapted from Kapitsa *et al.*, 1996).

It can be seen in Figure 8.7 that plotting experimental points determined from surface contours against "A-scope"-based ice thickness data taken on flight number 130 approximates a straight line, with the slope of the line close to that for freshwater or water of low salinity.

The same process was used to determine an "open" water level, or the level that the water would reach up a borehole drilled through the ice and into the lake itself. The level calculated was 3,100 m above sea level, an important result because it means the water of Lake Vostok has no contact with the ocean.

These data were among those presented by Dr. Robin and me at the "Lake Vostok Workshop" at the SPRI on 22–24 May 1995. Dr. Heap, Director of the SPRI welcomed the participants and explained the background of the meeting (the workshop was initiated as a result of his letter of 30 March 1995).

The "Background for the Workshop" of "The Draft Report of the Workshop" states "The local organizing committee at the SPRI (Drs. Robin, Zotikov, and Heap) considered that, for the workshop to be effective, participation should be limited to about 15 persons. It was recognized that not all aspects of the contents of SCAR XXIII-12, paragraph 3 could be covered and Dr. Heap advised that the

Environmental Impact Assessment (EIA) under the Protocol would likely require a Comprehensive Environmental Evaluation (CEE). The successful outcome of the EIA procedure would depend on the strength of the scientific case for drilling through the ice to the lake and the underlying sediments. Drs. Robin and Zotikov considered that the workshop should focus on defining the "relevant scientific investigations" which, by improving the description of glacial and subglacial environments of the lake, would significantly reduce uncertainties about the probable nature of the lake water and underlying sediments.

Dr. Heap suggested that the aims of the workshop should be to provide advice on how best to meet the provisions of paragraph 1 of the SCAR recommendation, to consider the scientific case for drilling through the ice into the lake, to discuss briefly possible non-polluting drilling technologies, and to outline scientific investigations that would assist in reducing uncertainties about the lake and its environs."

The report on the workshop, written after the meeting by Drs. Dowdeswell and Vaughan (glaciology) and Drs. Wynn-Williams and Ellis-Evans (lake water and sediments), included the current state of knowledge at that time (i.e., the first evidence of the presence of Antarctic subglacial lakes that was inferred from observations of extensive areas of very flat ice surfaces around Vostok Station). Evidence from airborne radio echo sounding later showed areas of unusual continuous reflections from the bottom of some areas of the ice sheet in regions of thick ice, especially around ice divides.

A list of interested specialists was developed as a means of involving people with relevant backgrounds and experience: Barkov, N. I., Russia, ice core studies; Bentley, C. R., U.S.A., seismic sounding; Dowdeswell, J. A., U.K., glacial geomorphology; Kamb, B., U.S.A., ice dynamics; Kotlyakov, V. M., Russian National Committee, glaciology; Lorius, C., France, French National Committee, ice core studies; Petit, J. R., France, ice core studies; Rees, W. G., U.K., radio-echo sounding; Rutford, R. H., U.S. National Committee, glacial geology; Tabacco, I. E., Italy, radio-echo sounding; Walton, D. W. H., U.K., SCAR, SCAR Group of Specialists on Environmental Affairs and Conservation; Watanabe, O., Japan, glaciology; and Wynn-Williams, D. D., U.K., microbiology. Many others expressed an interest in participating, so the original list of participants grew considerably. The sole participant from the U.S.A. was the president of SCAR.

The workshop agenda included discussions of four main sources of existing data for the Vostok area: (1) seismic reflection sounding, (2) airborne radio-echo sounding, (3) satellite radar altimetry, and (4) geophysical studies in the Vostok borehole. Highlights of the above areas of discussion are as follows:

(1) Seismic reflection sounding. Only two vertical seismograms survive from the fieldwork of the 1960s. Both will be interpreted to indicate a strong ice–water reflector at 3,700 m with a 500 m deep water column beneath. This is in turn underlain by what is interpreted as a 180–200 m thick accumulation of sediments, and then bedrock.

(2) Airborne radio-echo sounding. A flight along the long axis of Lake Vostok shows that the roof of the lake ranges from 3,700 m near Vostok Station to 4,200 m over a

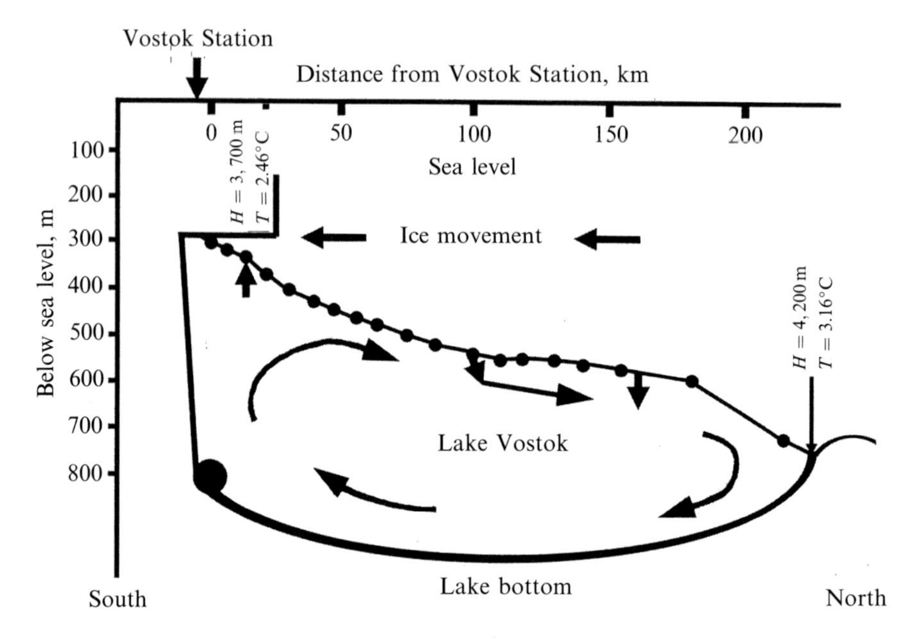

Figure 8.8. First cross section of Lake Vostok presented by Dr. Zotikov for the second (1995) workshop on subglacial lake exploration. The bottom of the ice cover of the lake is plotted using ice thickness data from "A-scope" picture studies; the lake's bottom position is plotted using only one data point (Kapitsa's seismic measurements at Vostok Station), the rest of the bottom is plotted hypothetically and later measurements have shown that the position of the bottom in this picture is wrong. Bent arrows are plotted to show the direction of flow of the main currents in the lake (these may need adjustment after future analysis).

distance of 250 km. This means that the melting point of the basal ice ranges between −2.4°C and −3.15°C. This may in turn generate water circulation within the sub-ice cavity (an ice pump mechanism) of Lake Vostok, with a configuration of an upper boundary based on 'A-scope' radio-echo sounding data. Figure 8.8 shows the ice shelf effects in the water cavern of Lake Vostok (the bottom of the lake is imaginary, represented by only one data point, at Vostok Station). Arrow directions of the lake water currents were also questionable at the time of this workshop. "In addition, the density of the lake water has been estimated from radio-echo derived ice thickness and the ice surface elevations from ERS-1 radar altimetry (Figure 8.7). The calculations suggest the water is unlikely to be saline. These data also imply that the ice above most of the lake is floating in hydrostatic equilibrium with the possible exception of a few local grounding points. The internal layers within the ice above Lake Vostok are also being investigated (Figure 8.9), as they contain significant information concerning ice dynamics and deformation in the region."

(3) Satellite radar altimetry. A 13 GHz satellite radar altimetry from ERS-1 has been used to define the ice surface topography over Lake Vostok and its drainage

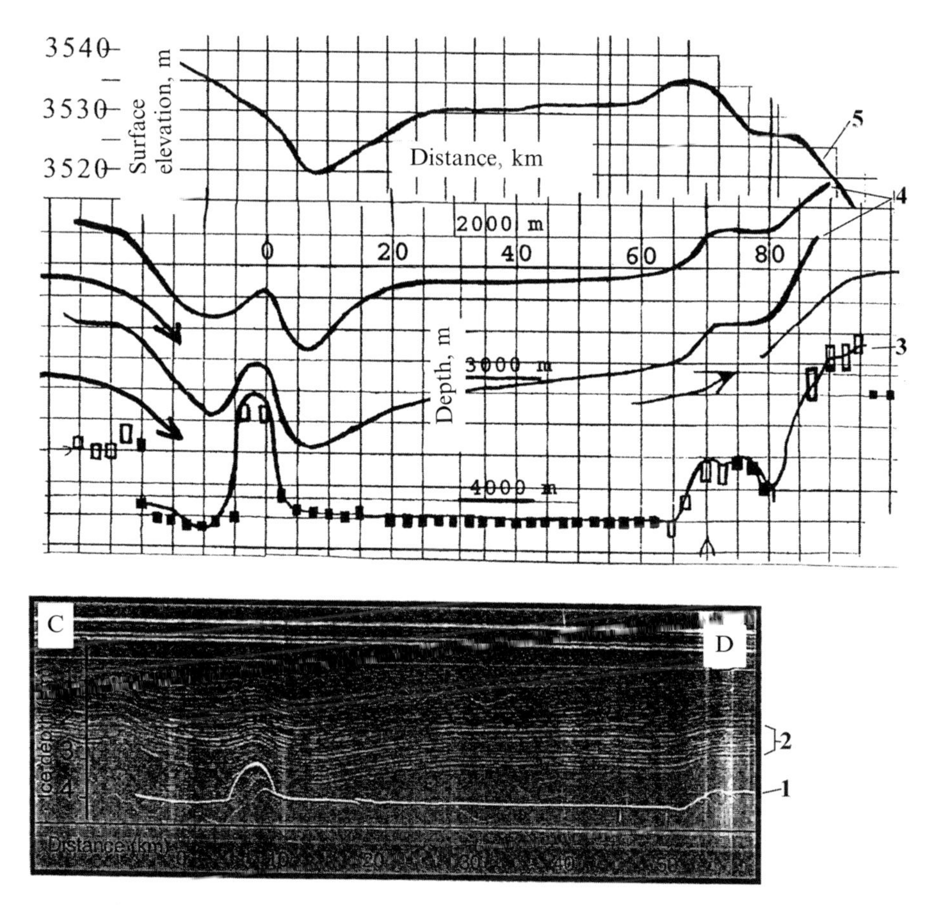

Figure 8.9. Cross section of a middle part of Lake Vostok along a line of the second part of flight number 130 (line C–D in Figure 8.1). (1) Bottom of Lake Vostok ice cover from a continuous record of bottom radio-echo reflections; (2) internal layers from a continuous record of internal radio-echo reflections; (3) bottom of Lake Vostok ice cover based on "A-scope" picture analysis (black square dots are the positions of an "A-scope" ice–water interface along the lake, open squares represent positions of "A-scope" ice–bedrock (no water) interfaces – the height of the squares indicates the uncertainty of the interface positions due to the width of interface reflections); (4) two strongest internal layers from a continuous record of internal radio-echo reflections; and (5) surface elevation. Note that "A-scope" pictures show that there is no water below the ice at the top of the "island"-type feature at the left and right sides of the curve of the ice bottom boundary. Distance along lines C–D is in kilometers from the top of left (west) "island". Arrows show the direction of ice flow.

basin to a very high accuracy. Accuracy of the ice topography is now ±1 m, with a precision of ±0.2 m, and is available every 350 m along a satellite track. Individual heights represent an average over a nominal 3 km footprint. These data have been used in the definition of lake extent and in calculations of lake water density. Depressions on the order of 5 m appear to define the inland margin of the lake, while a semi-continuous rise up to 12 m is associated with the downslope margin (Figure 8.6).

(4) Geophysical studies in the Vostok borehole. Since 1980 three cores have been drilled close to Vostok Station. Present ice flow models suggest that the ice at 2,000 m depth in the most recent core fell as snow 200 km upstream. The core represents more than 250,000 years of proxy climatic records. The Holocene, the last glacial with its stadial and interstadial episodes, a warm and relatively stable Eemian interglacial, and at least one further glacial–interglacial cycle is evident in the oxygen isotope, electrical conductivity, and other chemical signals. Methane and CO_2 gas concentrations also vary over glacial–interglacial intervals. Borehole closure studies yielded a value of 3.2 for the power flow constant of ice. The current borehole temperature profile is important for calculating the temperature at the ice–lake interface. Microbial analysis of the Vostok ice core has also been undertaken.

Other subglacial lakes

It was mentioned that more than 370,000 km of airborne radio-echo sounding profiles have been analyzed in order to look for information on subglacial lakes. More than 70 other subglacial lakes below the Antarctic Ice Sheet were found up to 50 km in length. Most of these lakes are located near ice sheet divides in the thickest parts of the East Antarctic Ice Sheet. It was reported that locations of these lakes are being compared with the ERS-1 radar altimetry data, and lakes with surface expressions larger than 10 km will be included on the map of Antarctica.

A summary of glaciological information of Lake Vostok was established at the workshop, as follows: location, 78.3°S–78.5°S, 102°E–106°E; area, 10,000 km^2; length, 230 km; maximum width, 50 km; ice thickness, 3,700–4,200 m; depth of water, 500 m (from single data point, Vostok Station); possible thickness of sub-lake sediments, 100–200 m (from single data point, Vostok Station); ice velocity, 3 m yr^{-1} (from single data point, Vostok Station); and surface accumulation rate, 2.3 g cm^{-2}.

Workshop participants also discussed other projects of deep drilling in central Antarctica. Deep drilling at Dome C is planned by The European Project for Ice Coring in Antarctica (EPICA). Deep drilling at Dome Fuji (Dome F) by the Japanese Antarctic Expedition was also discussed at the workshop.

Future glaciological requirements for Lake Vostok studies as identified in 1995 by the members of the workshop included, in order of priority. (1) Continuation of deep ice core drilling and maintaining an operational borehole status. (2) To make a simple vertical profiling experiment using an existing borehole. The aim is to get a new, high-accuracy seismic record of ice thickness, lake water depth, and sedimentation layer thickness, if it exists. (3) Further radio-echo sounding of the lake to provide more comprehensive data. (4) To measure surface velocity and strain rate if possible at one or more sites at Lake Vostok or its vicinity using a synthetic aperture radar (SAR) interferometer (as has been undertaken for the Rutford Ice

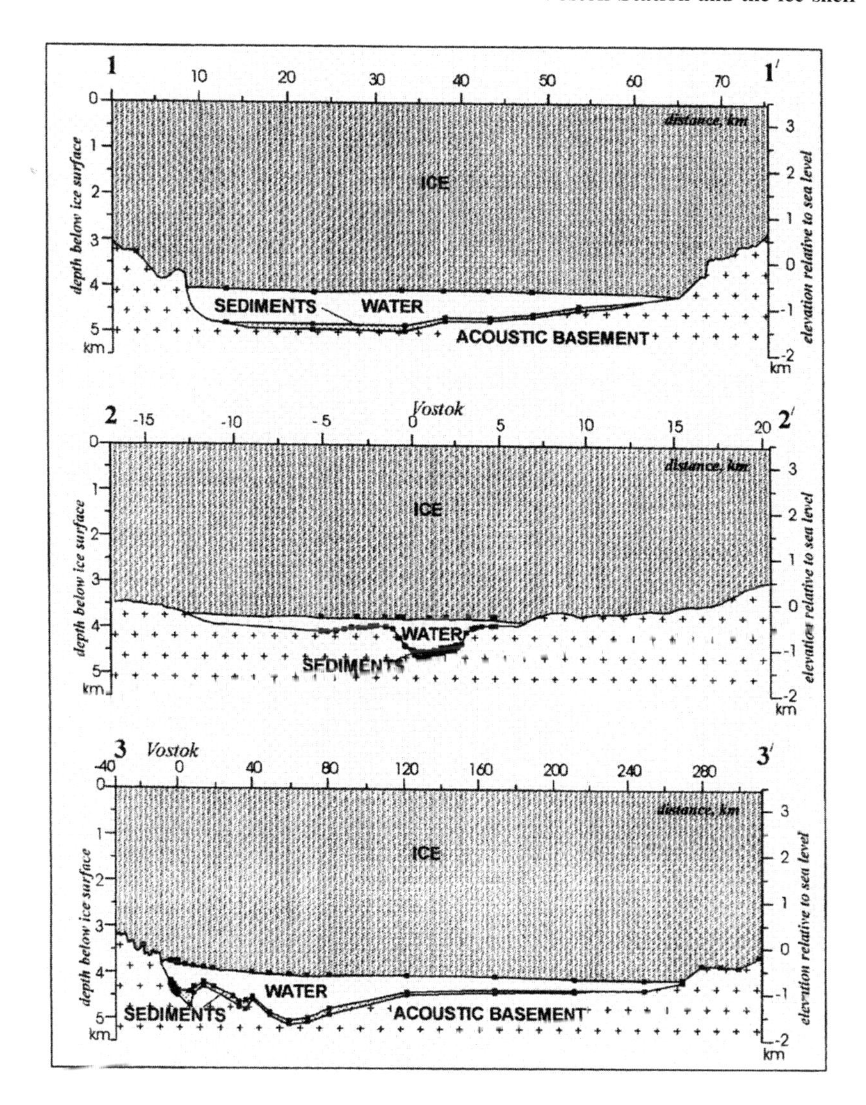

Figure 8.10. Combined radio-echo sounding and seismic measurements along the long axes of Lake Vostok (3–3′ in Figure 9.1) and two transverse profiles across the lake (1–1′ and 2–2′ in Figure 9.1) performed in 1995/1996 austral summer by the RAE. The expedition revealed the actual position of the lake bottom and the existence and size of the bottom sediment layer (adapted from Masolov *et al.*, 1999).

Stream) and/or global positioning system (GPS) methods with the aim of obtaining an ice velocity field and flow regime.

Results of this discussion initiated an immediate reaction and combined radio-echo sounding and seismic measurements were made along the long axes of Lake Vostok and two transverse profiles across the lake in the 1995/1996 austral summer by the Russian Antarctic Expedition (RAE) (Figure 8.10). They revealed a more accurate position of the Lake bottom and existence and dimensions of the bottom sediments (Masolov *et al.*, 1999).

An interesting discussion occurred about the lake water and bottom sediments. It was postulated that a radio-echo radar can penetrate freshwater through depths of 15–25 m and in any circumstances up to 50 m. It was agreed that Lake Vostok is a relatively freshwater lake. The meeting also accepted the possibility of some form of circulation of lake water because of a gradient of about 500 m elevation in the ice roof of the lake from its northern to southern end, giving a thermal gradient of 0.50°C due to the pressure dependence of the water's freezing temperature, which changes from −3.15°C at the ice–water boundary at the northern end to −2.45°C at the southern end. Water convection due to this temperature difference may be "as little as 1–2 m yr^{-1}, but this is enough to have a dramatic effect on the ice–water system in terms of heat flow". Following this approach a few years later, scientists developed a computer model of the water circulation processes in Lake Vostok below the Lake Vostok Ice Shelf, which showed a circulation of the kind in Figure 8.8 for slightly salty water and a circulation of opposite direction for pure water (Figure 8.11).

The concluding words of the Lake Vostok workshop in Cambridge, 22–24 May 1995, were also words of warning, stating that local heat transfer conditions at the ice–water interface could be different in areas of the lake, and an area of localized bottom freezing found below Vostok Station is a product of that origin.

The workshop also initiated a new, more serious approach to the biological study of the lake. It was first considered that the only life that would be found in the lake or in its sediments would be microbial (bacteria). There would not be enough nutrient material in the lake to support any living bacteria and that no known bacteria would be able to exist for a long period in water in a state of suspended animation. There would be no equivalent of a Loch Ness Monster here. However, it was recognized that there could well be bacterial flora contained in the lowest ice layers that could be reactivated when lake water enters the base of the borehole and migrates into the ice wall. It was also suggested that the moving ice could carry nutrient carbon into the lake from coal and hydrocarbon deposits upstream and from sources of volcanic activity, but such carbon would not be in a form suitable for bacterial nutrition.

Remarks of the workshop on the importance of the micro-organisms in ice

Microbes are important because of their capacity for logarithmic growth in conditions suitable for them. One bacillus could provide a mass equal to that of the Earth in living material within 36 hours if unlimited food and space was available and accumulated toxins were removed. Favorable conditions can be very undemanding because microbes can grow in very difficult conditions. Diverse microbes isolated from ice cores are a genetic resource whose DNA can be amplified by genetic engineering using a polymerase chain reaction (PCR). This gene pool is not only a valuable scientific tool for describing new microbes, microbial history, and community structure, but also a resource for potentially valuable materials of

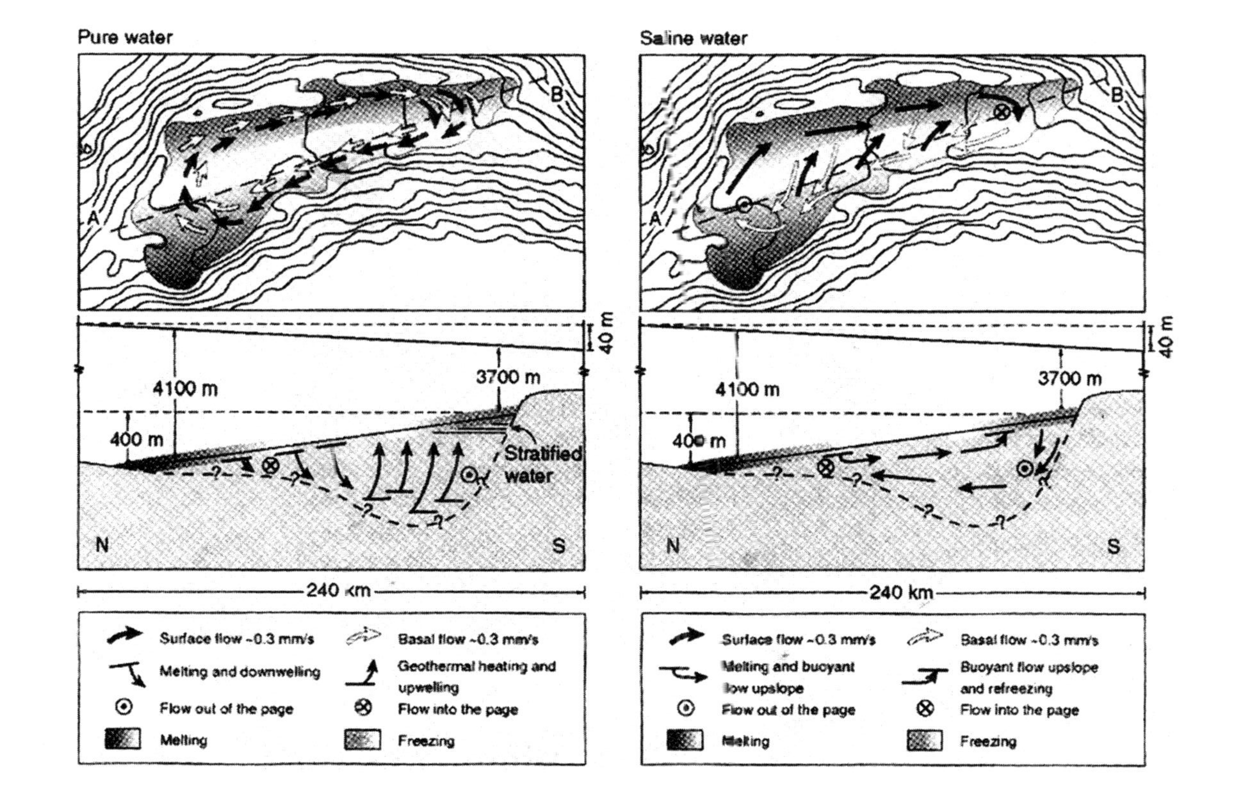

Figure 8.11. Water circulation within Lake Vostok below the Lake Vostok Ice Shelf and above and still uncorrected lake bottom topography. Left-hand side: circulation calculated by numerical modeling, assuming the water is pure (after Williams, 2001; Wuest and Carmack, 2000). The white arrows show the bottom water circulation and the black arrows denote a higher level circulation close to the base of the ice. Dots refer to upwelling of lake water, crosses denote downwelling. There are two clockwise circulation paths in the upper and lower areas of the lake. Most of the vertical mixing takes place in the southern two-thirds of the cavity, but this exchange is rather limited. Right-hand side: the lake is saline to a small extent (Souchez *et al.*, 2000) (adapted from Siegert *et al.*, 2003).

medical and economic importance. Ice and sediment microbes are sensitive indicators of climate change, such as that reported by Abizov (2001) representing an 8°C change at approximately 200,000 years. It is important to mention that these microbes represent the micro-biota existing before major global pollution processes started.

"Micro-organisms found to date in the youngest layers of ice at Vostok by Dr. Abizov include non-spore forming bacteria of well known genera (Abizov, 1993), but only about 1% of the total micro-biota can be grown under currently known culture conditions. There is therefore much more genetic potential if PCR techniques are used. Yeasts have been isolated from ice formed 3,000 years ago and filamentous fungi formed 36,600 years ago. Actinomycetes, branched bacteria, renowned for antibiotic production, survive for 2,200 years and include a new species. A unicellular alga has been isolated from 119,000 year old ice, whilst dormant spore-forming material can survive for at least 200,000 years (at 2,395 m). Their value for science includes fundamental studies of their membrane structure and function, their ultra structural adaptations and their survival potential. This concerns their tolerance of low temperatures, high pressures, high local osmotic stresses, nutrient starvation and high oxygen concentrations (hyperoxia) due to exclusion of gases from re-freezing water" (From Drs. Wynn-Williams and Ellis-Evans, during the Lake Vostok workshop).

In the absence of free water (at about −55°C in the upper part of the ice sheet) microbial survival is extremely long due to the protection of critical molecules such as proteins from osmotic stress by compatible solutes (antifreezes). Bacterial viability in the pure ice of the Vostok area is not as great as in the Siberian permafrost (up to 3 million years) because of the protective function of mineral particles that concentrate nutrients and permit mutual support within biofilms (Gilichinsky *et al.*, 1995). Morainic inclusions in Vostok ice may fulfill this function. In Antarctic habitats, such as endolithic communities within rocks, viability times can be up to 23,000 years, so that growth rates in low-temperature ice may be very slow. Long-term radiation damage to microbial DNA has been reported in permafrost environments and requires basal metabolism for its repair. Spores of bacteria and other microbes, including yeasts and filamentous fungi, have been shown to remain viable for between 25 and 135 million years in the intestines of bees preserved in amber (Cano and Borucki, 1995), although their carbon may have been recycled in this closed-circuit habitat. This indicates a potential for the viability of quite old microbes in the remaining ice profile above Lake Vostok down to 3,600–4,200 m. The transition zone above the lake, where the pressure melts ice and ice crystals with free water is an area of great microbiological interest. Organisms in this zone of potential accumulation of nutrients and active metabolism at low temperature will inoculate the oligotrophic (low nutrient) lake water and initiate a microbial sedimentation process. Study of this microbiota requires strict aseptic procedures, including controls such as the use of minute bacteria such as Serratia Marcescents to test for the penetration of the ice fractures and contamination of equipment by drill fluid (Abizov, 1993, 2001).

Main biological reasons for penetrating the lake ***(from the May 1995 workshop)***

The lake has been isolated from the atmosphere for tens or hundreds of thousands of years and is therefore unique. It is not known if the lake is derived from the pre-glacial lake or a system formed by glacial processes. A pre-glacial lake may have left some evidence (even cultivated material) within the lowest sedimentary strata. This would likely be a more diverse fossil flora and fauna than the modern lake contains. It is likely that viable organisms would not survive from a "fossil lake", but nucleic acids might be present that would respond to the PCR technique.

The water column itself is unlikely to contain significant numbers of microbes because of its extremely nutrient-poor status and limited mixing. The presence of free water will enforce microbial growth and immediately stress those organisms released from the ice. The lake presents a number of significant stresses. First, the low temperature (around −2°C) will be unacceptable for methophiles. The pressure imposed by 4 km of ice will equate to that which is found in the average ocean depth and many micro-organisms deposited and frozen into the ice cover will not be tolerant of such pressures. The nutrient status of the water column will impose an immediate severe stress on most cells and there is a potential for a marked osmotic change on being released from the ice. Dragging of water by the ice sheet towards the downstream lake edge will result in ice formation at that point, and thus exclusion of dissolved gases, notably oxygen and nitrogen. Similar gases trapped and compressed in the ice sheet will have been constantly released on melting of the under-ice sheet and contribute to a potentially hypoxic state. High oxygen concentrations are lethal to many organisms but are found in several perennially ice-covered Antarctic lakes. Those organisms which can survive the physiological shock of release from the ice transition zone will probably exist either as dormant spores or as stressed micro-cells.

The main area of potential microbiological interest is the sediment that may be as much as 200 m thick and is probably continually replenished by sedimentation from the ice and, quantitatively more significantly, by lateral deposition of morainic debris. The morainic material will be inorganic with negligible organic content. The presence of mineral particles to act as potential substrata for microbial biofilms is significant.

The continual inoculation of the sediment by deposition from the ice increases the statistical probability of viable forms and it seems highly probable that despite the stresses, viable micro-organisms will occur. Hot spots of geothermal activity might permit significant local community development in a habitat isolated from current climatic changes and man-made environmental impacts. The potential population would have a valuable gene pool and for the ice core microbiota, with all the associated benefits for science, medicine, and commerce.

The use of PCR techniques on the DNA of the organisms would yield valuable information on the history and biochemical potential of an ancient microbiota. There may also be implications for exobiological studies which also concern life at extremes.

Carbon supplies will be very limited and, in circumstances of low energy status, much of the carbon may be unavailable as it takes too much energy to break down and utilize. Knowledge of the community structure would help to resolve the origin of the lake either as a "fossil lake" or as a primordial sub-ice water body. It may also address the issue of the possible catastrophic deglaciation of the Pleistocene Ice Sheet.

The sediment microbiota will reflect the dynamic nature of the lake in terms of morainic deposition, biotic and mineral sedimentation, temperature gradients, convection, refreezing, and geothermal activity. If the lake can be penetrated with a minimal contamination (some environmental impact is inevitable but can be minimized by careful planning), much valuable microbiological information could be derived from sediment cores if not from the water column.

Summary of selected recommendations *(from the May 1995 workshop)*

Workshop participants reviewed the evidence for the existence of a large body of water beneath the inland ice in the area of Vostok Station and extending beneath it. They agreed the following recommendations.

(1) Presently ongoing core drilling at Vostok Station should proceed, but not penetrate into the water beneath the ice. Termination of drilling can be determined while drilling from measurements of ice temperature and remaining ice thickness. Based upon our present state of knowledge the minimum remaining ice thickness should not be less than 25 meters.
(2) Additional geophysical surveys should be performed to better define the glaciological and geological setting of the lake. The highest priority is to be given to seismic measurements in order to confirm the existence of a large body of water and underlying sediments.
(3) Provided the existence of a large water body can be confirmed, studies must be made to find techniques for accessing and sampling the water and sediments beneath with minimum contamination of this environment. Consideration should be given to testing such techniques at alternate sites, where monitoring would be possible.

These recommendations are published here to stress that participants of this workshop, aware of all the information about the lake's discovery, avoided admitting that the Lake exists, but instead left it to a scientific community to determine the existence of the lake. We know now that the existence is widely and enthusiastically endorsed.

9

New data on Lake Vostok

Further data on Lake Vostok resulted from two meetings following that at the Scott Polar Research Institute (SPRI) in 1995: (1) the International Workshop on "Lake Vostok Study: Scientific Objectives and Technological Requirements", 24–26 March 1998 at the Arctic and Antarctic Research Institute (AARI) in St. Petersburg, Russia, and (2) the "SCAR International Workshop on Subglacial Lake Exploration" held in Cambridge, England, September 1999.

The meeting in St. Petersburg was intended to invite as many of the scientists from AARI as possible, some of whom had been involved in studies of the Vostok Station area for 41 years, since 1957. Scientists from St. Petersburg's State Mining Institute were also invited. That institute had been in charge of deep-core drilling through the ice sheet at Vostok Station for 26 years since 1971. It was this drill, operating at a depth of 3,623 m from a drilling rig at Vostok Station, that reached a depth only 130 m from the lake itself on the opening day of the workshop. Other Russian scientific institutions involved in the studies of Vostok ice cores and the lake itself were also present. One was the Institute of Microbiology of the Russian Academy of Sciences, represented by Dr. Abizov, who identified micro-organisms deep in the ice sheet below Vostok Station and who had been involved in its study for more than 20 years. St. Petersburg's Polar Marine Geological Research Expedition had been involved in remote sensing of the Vostok area for the last few years.

After a welcome and introduction by the Director and Deputy Director of the AARI, a presentation by the Leader of the Russian Antarctic Expedition (RAE), Dr. Lukin, was made on the subject of "Russian Antarctic studies at Vostok Station". The history of the discovery of the lake and possible means of further exploration were discussed, and new data was presented from new studies since the lake became a public issue.

The SCAR Workshop in 1999 was supported by the British Antarctic Survey (BAS), the U.S. National Science Foundation (NSF), the European Science

Foundation (ESF), the British Council and the U.S. National Aeronautics and Space Administration's (NASA) Astrobiology Program.

The meeting in Cambridge was convened by Dr. Ellis-Evans at the request of the SCAR Executive and its President, Professor Rutford, and this choice determined the biological concern of the meeting in general.

Scientists from the AARI, St. Petersburg's State Mining Institute, Laboratoire de Glaciologie et de Géophysique de l'Environnement, Grenoble, France, and Laboratoire de Modélisation du Climat et de l'Environnement, Saclay, France, reported that under the framework of Russian–U.S.A.–French collaboration on the Vostok ice core study, drilling operations at Vostok Station's "5G" borehole had proceeded after the Cambridge workshop of 1995. During the 1996–1997 and 1997–1998 field seasons the hole was deepened from 3,350 m to 3,623 m. A core from the lake's vicinity was taken through this depth also, and preliminary studies were made (Petit *et al.*, 1998). It was shown that from the surface of the ice sheet down to 3,300 m depth, stable isotope, dust, and electrical conductivity measurements (ECM) are well preserved in the ice, offering a continuous climate record over the last 400,000 years. However, below 3,400 m the isotopes and ECM signals became smooth and it was no longer possible to decipher glacial–interglacial changes. This meant that two decades of drilling and the associated climatic studies of the Vostok core came to an end. The core has the unique distinction of providing access to the most extended undisturbed paleo-environmental series covering 420,000 years, including the last four climatic cycles (Petit *et al.*, 1997). Rather than terminating the studies, however, new and interesting information began to appear. A very sharp and significant variation in stable isotope and dust concentrations around 3,320 m depth were found, which could not be of climatic origin. The ice stratum from 3,538 to 3,607 m depth displayed visible inclusions of millimeter size, which might have originated from bedrock. The layer also included very large crystals, meters in size (the diameter of the core was about 10 cm, its length was about 6 m – these large ice crystals could therefore only be measured along the length of the core). Electrical conductivity of this layer was two orders of magnitude lower than previous layers, but the stable isotope content was only slightly different. It was clear that a new area of the ice mass had been reached, but what did it represent? In answer to this question, the presentation by AARI scientists on the internal structure of the lower part of the Vostok ice core was of great interest at the time (Lipenkov and Barkov, 1998).

> *The basal stratum, found beneath 3,538 m and traced downward to the bottom of the hole (3,623 m), is represented by silt-sized material comprising randomly disseminated moraine debris. The very bottom part (3,606–3,623 m) of the stratum appears to consist of relatively clean ice. The large crystal size (as much as 1 m or more) is of little help to determine the ice texture pattern (fabric). The presence of inner moraine in the basal stratum at Vostok implies that there is a probability for refrozen ice to be found at a distance of more than 100 m from the bottom of the ice sheet.*

The authors of this presentation would not know until further studies took place in a few years that this area consisted of a layer of ice formed by bottom freezing of the lake's ice. The conclusion of the presentation stated that "continuation of ice coring in hole 5G-1 down to the terminal depth permitted by the Scientific Committee on Antarctic Research (SCAR) (3,700 m) using the same equipment should be proposed as the next reasonable step toward Lake Vostok discovery."

It is interesting that in 2004, only 6 years after the workshop, the majority of scientists agreed that this bottom layer was formed by lake water freezing. Calculations of the scientific explanation of the possibility of freezing at the ice sheet–lake water interface below Vostok Station were also presented at this workshop (Salamatin, 1998).

In the case that there was no shear deformation in the ice sheet above the lake, and strain heating in the glacier was negligible, the best fit between the computed and measured borehole temperatures at Vostok was found to be $0.043\,W\,m^{-2}$ for the bottom heat flux, resulting in a mean water freezing of $0.75\,mm\,yr^{-1}$. This could result in a 20–30 m thick layer of newly frozen ice at the ice sheet bottom in the vicinity below Vostok Station.

Three years (1996–1998) of seismic studies during St. Petersburg's Polar Marine Research Expedition by AARI in the vicinity of Vostok Station along the submeridional and sublatitudinal profiles, and in the borehole itself, have shown that the ice sheet thickness at the location of borehole 5G-1 is 3,750 m and there is an underlying water layer with a thickness (depth) of 670 m. The locations of remote-sensing data in the Lake Vostok area taken from 1959–2000, parts of which were discussed at these workshops, are shown in Figure 9.1. Discussions at the St. Petersburg workshop in 1998 resulted in new plans for future radio-echo sounding and seismic investigations (Figure 9.2).

One of the reports of the workshop included a discussion on the tectonic setting of Lake Vostok (Leitchenkov *et al.*, 1998). The lake is located at the edge of a large area of bedrock beneath the thick ice sheet of East Antarctica, which is interpreted as a Precambrian crystalline shield. Bedrock topography and geophysical data available for this region suggest that Lake Vostok is associated with an intercontinental rift zone similar to those of other regions on Earth (e.g., rifts of East Africa, the Baikal Rift, the St. Lawrence Rift, and others) (Figures 9.3 and 9.4).

It should be emphasized here that the main borehole at Vostok Station, and the spot of the first seismic sounding by Kapitsa are located surprisingly close to the shore of the lake. The coastline of Lake Vostok is positioned at a distance of only 3 km west, 4 km east, and 4 km south-westward of borehole 5G-1, which are of nearly 1% of the lake's length.

The measured ice thickness to the north of Vostok Station along a submeridional 6 km long profile was the same as for borehole 5G-1, about 3,750 m. The distance from the bottom of the hole and the ice–water boundary is 130 m (Popkov *et al.*, 1998).

The water layer under the ice was measured remotely, showing a change in thickness (depth) from 490 m at the southern part of the profile to 670 m at its northern part.

Additional investigative scientists appeared at this stage: in biology, Drs. Abizov

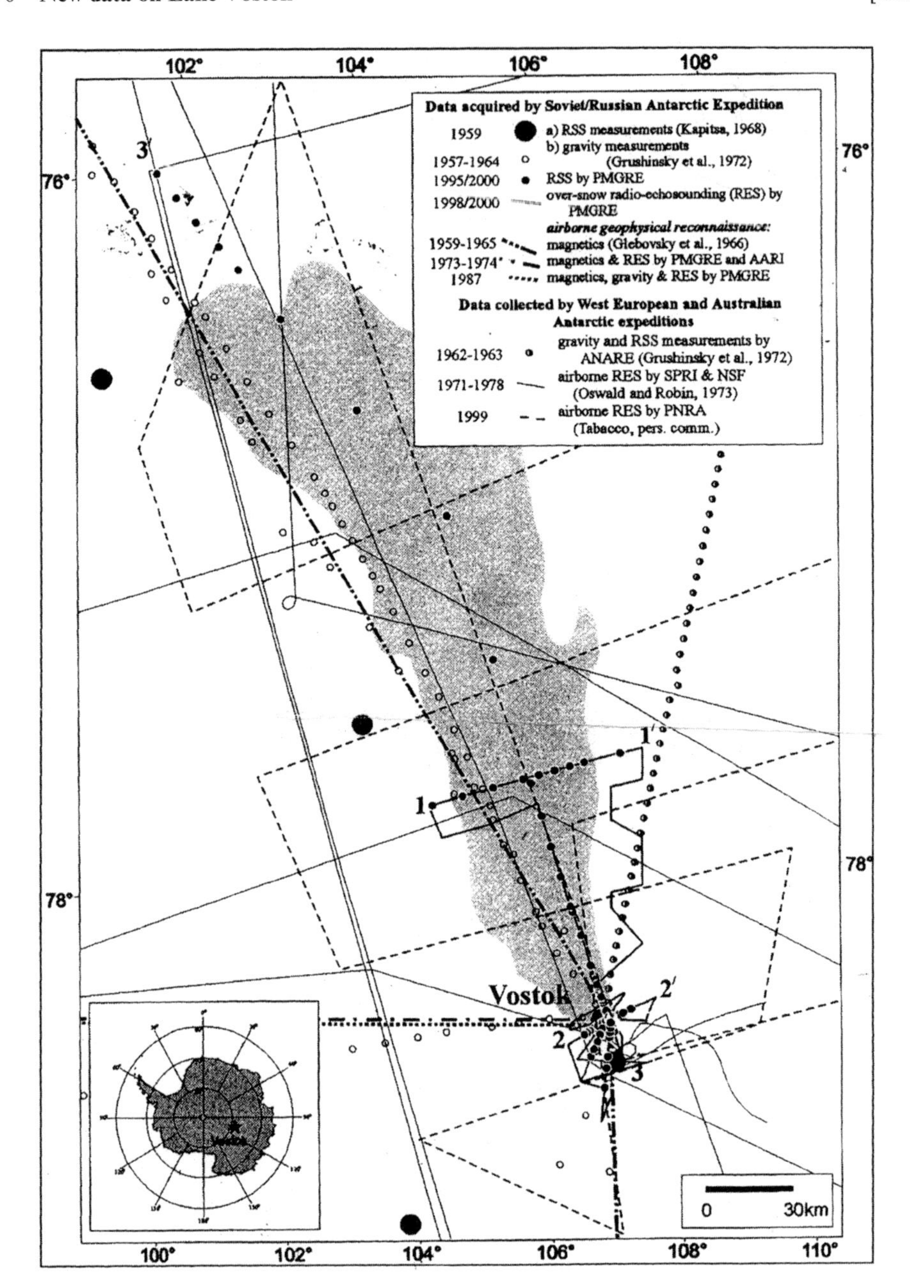

Figure 9.1. Location of remote-sensing data in the Vostok area from 1959–2000. The location of Lake Vostok is shaded (adapted from Masolov *et al.*, 1999). Notation: RSS – reflection seismic soundings; RES – radio-echo sounding; PMGRE – Polar Marine Geological Research Expedition (Russia); AARI – Arctic and Antarctic Research Institute (Russia); ANARE – Australian National Antarctic Research Expeditions; SPRI – Scott Polar Research Institute (England); NSF – National Science Foundation (U.S.A.).

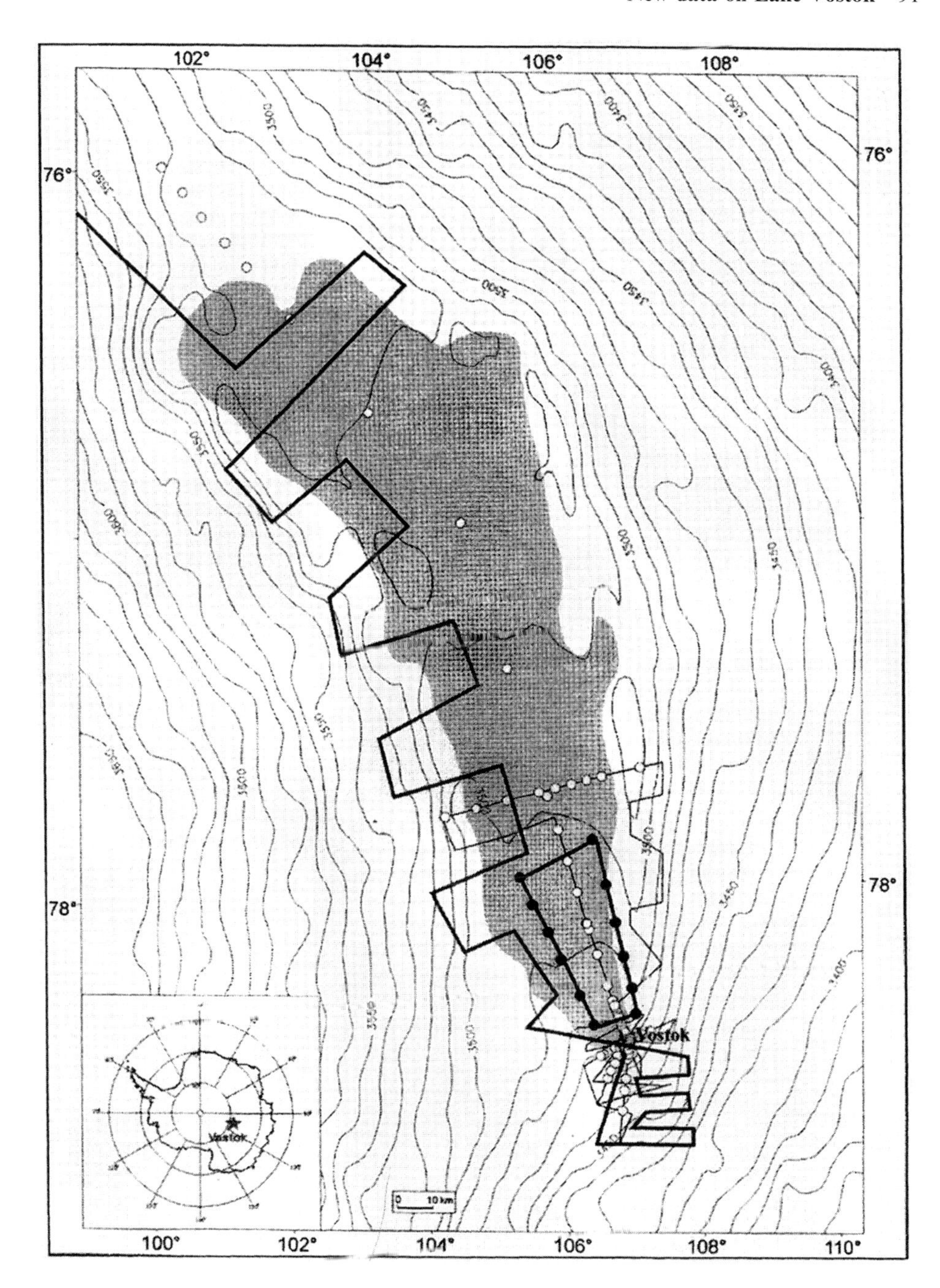

Figure 9.2. Over-snow radio-echo sounding (thick lines) and reflection seismic sounding (dots) investigations planned by the Polar Marine Geological Research Expedition as a part of the Russian Antarctic Expedition for 2000/2001. Thin lines and open circles indicate earlier surveys. Ice surface elevation contours are in meters (adapted from Masolov *et al.*, 2001).

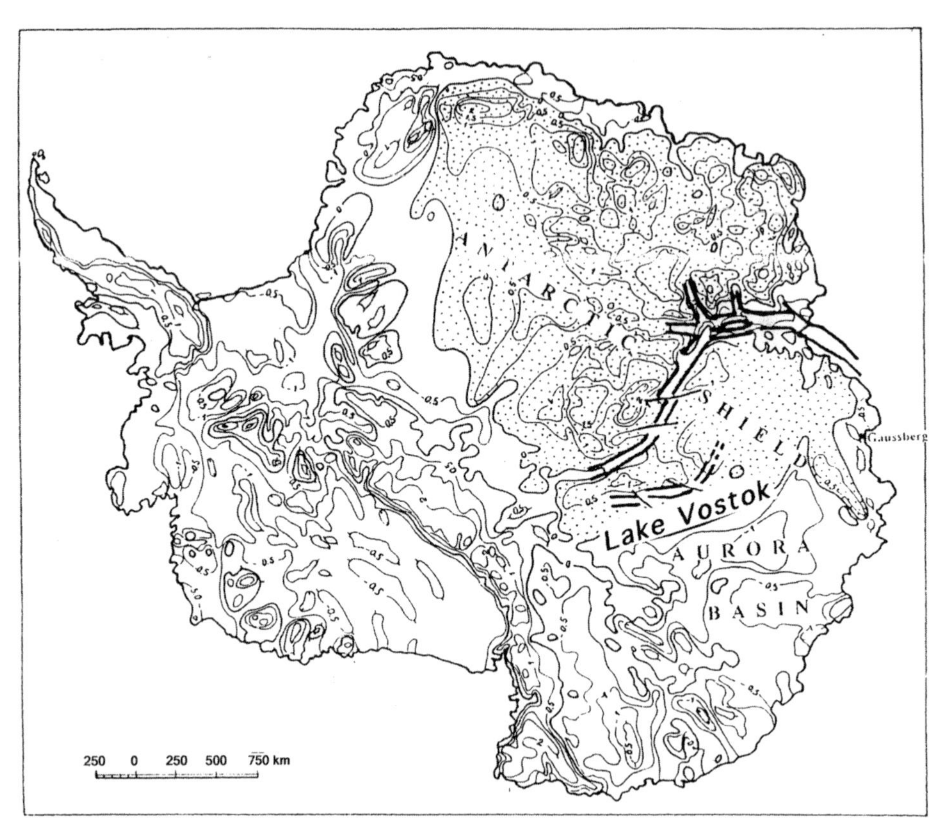

Figure 9.3. Location of the Lake Vostok Rift in the tectonic structure of Antarctica (adapted from Leitchenkov *et al.*, 1998). The stippled area represents the inferred Antarctic Precambrian Shield; thin lines – bedrock topography contours; thick lines – rift zone boundaries.

and Vorobieva from Russia, Dr. Ellis-Evans from England, and Dr. Doran from the U.S.A. Dr. Abizov summarized his years of studies of microbes from the deep Vostok core (Abizov, 2001) thus:

> *Will we find unique organisms in Lake Vostok? asked Dr. Ellis-Evans from the British Antarctic Survey (Ellis-Evans, 1998) ... Antarctic lakes have yielded a full range of microbial form and function despite the relatively few studies undertaken to date. Work in the Vestfold Hills revealed a unique protozoan species surrounded by a halo of bacteria connected by fine filaments to the protozoan. This assemblage interacts metabolically, presumably for mutual benefit and exists only near the oxic/anoxic interface. The evidence from studies of Antarctic lakes suggests that ecophysiological adaptations will occur in microbial communities. Substrate availability will be a major issue in Lake Vostok, as in other Antarctic lakes and physical surfaces will be where*

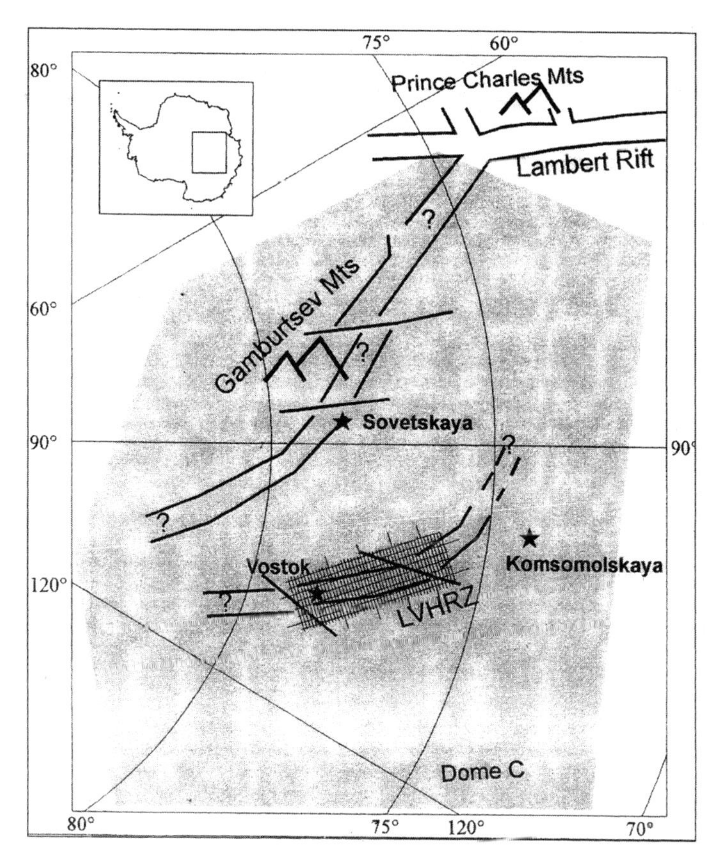

Figure 9.4. Aerogeophysical investigations planned by the U.S. Antarctic Research Program (USARP) for 2000/2001 (thin lines) and the proposed area for investigation by the Polar Marine Geological Research Expedition (Russia) for subsequent years (shaded) (adapted from Masolov *et al.*, 1999). Notation: LVHRZ – Lake Vostok Hypothetical Rift Zone. (Location of USARP flight lines courtesy of Dr. Bell).

nutrients are likely to concentrate. Assuming some degree of mixing in the water column and a lack of nutrients the obvious area to search for microbes will be the sediments and the ice–water interface where surfaces abound and chemical gradients are likely. Nutrients are present in sediments but will require considerable energy outlay to become available. Is this energy available? Evidence from maritime Antarctic lakes isolated in both batch and chemical culture suggests that the majority of these isolates can tolerate exceptionally low levels of C and N. Most of the aquatic microbes are thought to live naturally in a state of long-term starvation broken by occasional 'feasts' ... Certain fundamental maintenance processes have to occur within a microbial cell for life to continue and this is an inescapable drain of cell resources. Bacterial spores have taken the survival process to an excep-

> *tional degree but still have limits, and spore-formers are not that common in Antarctica. [In contrast, Dr. Abizov has now shown that this is not the case for the deepest horizons of the Vostok ice core.] The reality in most low nutrient environments will be that lysed microbes provide much of the nutrient supply so that circling is tightly bound. Will this impose limitations on species diversity and will processing of micro-molecules be limited because of the high energy demands involved in these processes? In many cold aquatic environments, macro-molecule breakdown is the major limiting step in nutrient circling and this may be an even larger hurdle in Lake Vostok.*

Dr. Vorobieva and Dr. Gilichinsky (Soil Science Department of Moscow State University and Institute of Soil Science and Photosynthesis of RAS) raised an optimistic point at the workshop, emphasizing that a high abundance of microbial biomes in Antarctic permafrost conditions, presence of quick reparated cells and signs of cell stabilization, and preservation of taxonomical and physiological diversity of microbial communities are all indications of microbial adaptations to long-term cold (Vorobieva *et al.*, 1998). With that in mind, it is possible that mechanisms of microbial survival are universal in a wide range of water–gas–mineral systems, and optimism is increasing in the search for life beyond Earth.

Dr. Doran pointed out that deep ice-covered lakes in the dry valleys of Victoria Land should be viewed as aquatic analogs of the Lake Vostok environment. "Like the dry valley lakes the Lake Vostok ecosystem will be entirely microbial, and metabolism extremely slow. In both systems there is a water body accepting microbes and organic carbon from an overlying ice cover, even though there is no light penetration in the dry valley lakes or in Lake Vostok. The dry valley systems are archaic with long residence times. Atmospheric gases in Lake Vostok may be supersaturated as they are in the dry valley lakes..." (Doran, 1998).

A new branch of science, that could be called extraterrestrial studies, was included in the meeting – Drs. Carsey, Cutts, and Harvath from the Jet Propulsion Laboratory of the California Institute of Technology introduced a proposal to NASA and the U.S. NSF, as well as to the international community on robotic exploration of Lake Vostok. This exploration is perceived to be closely related to the deep subsurface exploration of other planetary sites, especially Europa, a moon of Jupiter, in the more distant future.

10

Plan for the international study of Lake Vostok

A question was raised with regard to scientific research related to Lake Vostok at the workshops mentioned in earlier chapters – namely, where were the American scientists who have pursued the traditional directions of glaciological and geophysical studies and who have spent time studying bottom processes of the ice sheet before 1996? Except for a few relatively new American glaciologists, little effort or research was spent by the U.S.A. on the issues of Lake Vostok. The answer lies behind their main interest of present study, that of the bottom processes of the West Antarctic Ice Sheet, including the study of the known subglacial ice streams that feed the Ross Ice Shelf. However, the interest of American scientists in subglacial lakes peaked with the possibilities of finding new (or old) forms of life.

Interest by the U.S.A. became apparent when a further workshop on Lake Vostok was announced for 7–8 November 1998 in Washington, D.C., only 6 months after the meeting in St. Petersburg and 11 months before the Scientific Committee on Antarctic Research (SCAR) Workshop on Subglacial Lake Exploration, in Cambridge. The main topic of the workshop, sponsored by the U.S. National Science Foundation (NSF), was evident in the title of the final report – "Lake Vostok: A Curiosity or a Focus for Interdisciplinary Study?" (Figure 10.1).

The workshop's conveners were Dr. R. Bell from the Lamont–Doherty Earth Observatory and Dr. D. Karl from the School of Ocean and Earth Science and Technology, University of Hawaii. Dr. Bell reviewed the geology and geophysical aspects of the problem, and Dr. Karl the microbiological component. Four small parts of pages 2–8 from the final report of this workshop are cited below.

Part 1 (Bell and Karl, 1998, pp. 2–3). Life continues to appear in the unusual and extreme locations from hot vents on the seafloor to ice-covered hypersaline lakes in Antarctica. The subglacial environment represents one of the most oligotrophic environments on earth, an environment with low nutrient levels and low standing stocks of viable organisms. It is also one of the least accessible habitats. Recently the

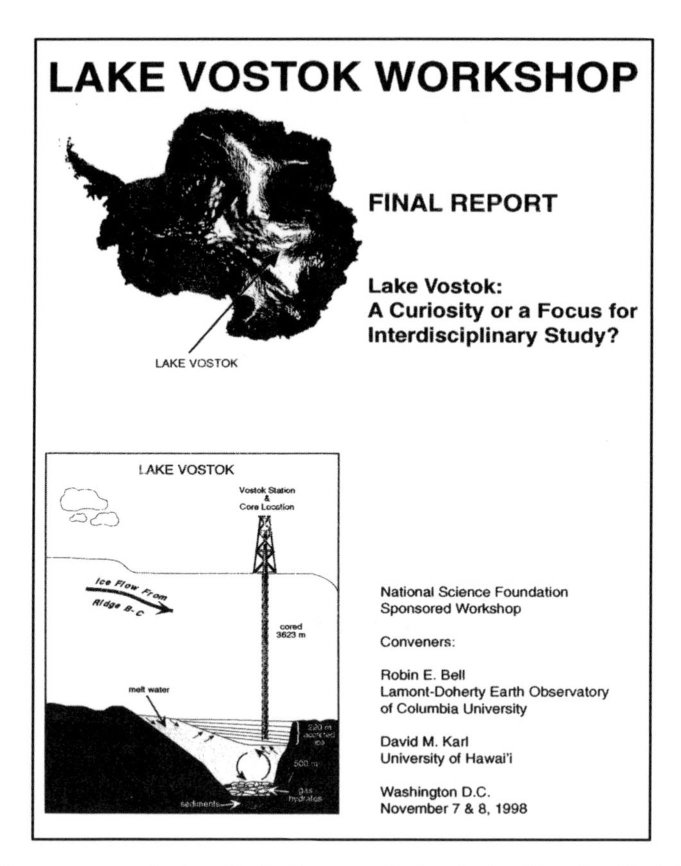
LAKE VOSTOK WORKSHOP

LAKE VOSTOK

FINAL REPORT

Lake Vostok:
A Curiosity or a Focus for Interdisciplinary Study?

LAKE VOSTOK

Vostok Station & Core Location

Ice Flow From Ridge B-C

cored 3623 m

melt water

National Science Foundation Sponsored Workshop

Conveners:

Robin E. Bell
Lamont-Doherty Earth Observatory of Columbia University

David M. Karl
University of Hawai'i

Washington D.C.
November 7 & 8, 1998

Figure 10.1. Front page of the Final Report of the Lake Vostok Workshop held in Washington, D.C., 1998. Sponsored by the National Science Foundation.

significance of understanding subglacial communities has been highlighted by discoveries including the thriving bacterial communities beneath alpine glaciers, to the evidence from African stratigraphy for a Neoproterozoic snowball Earth to the compelling ice images from Europa, the icy moon of Jupiter. If life thrives in these environments it may have to depend on alternative energy sources and survival strategies. Identifying these strategies will provide new insights into the energy balance of life.

The identification of significant subglacial bacterial action as well as the work on permafrost communities suggests that life can survive and possibly thrive at low temperatures. Neither the alpine subglacial environment nor the permafrost environment is as extreme as the environment found beneath a continent-wide ice sheet as Antarctica today. The alpine subglacial environment has a continual high level of flux of nutrients from surface crevasses. The Antarctic subglacial environment lacks a rapid flux of surface meltwater and subsequently is more isolated. In addition to being more isolated, the Antarctic subglacial environment is a high-pressure region due to the overburden of ice.

The Antarctic subglacial environment may be similar to the environment beneath the widespread ice sheets in the Neoproterozoic, a time period from about 750 to 543 million years ago. It has been suggested that during this period the Earth experienced a number of massive glaciations – covering much of the planet for approximately 10 million years at a time. The evidence for an ancient ice covered planet comes from thick widespread sedimentary sequences deposited at the base of large ice bodies. These glacial units alternate with thick carbonate units – warm, shallow-water sedimentary deposits. These paired sequences have been interpreted as representing a long period when the Earth alternated between an extremely cold, completely ice-covered planet (the snowball Earth) and a hothouse planet. Some speculate that the extremes of these climates introduced an intense "environmental filter", possibly linked to a metazoan radiation prior to the final glaciation and an Ediacaran radiation. Portions of the Antarctic continental subglacial environment today, which have been isolated from free exchange with the atmosphere for at least 10 million years, are similar to the environment in this ancient global environment. Understanding the environmental stresses and the response of the microbes in a modern extreme subglacial environment will help us decipher the processes which lead to the post-glacial evolutionary radiation over 500 million years ago.

The third important analog for modern Antarctic subglacial environments is from the outer reaches of the Solar System, the ice moon of Jupiter, Europa. Recent images resembling sea ice, combined with the very high albedo of this moon have led to the interpretation that this moon is ice covered. Beneath the ice covering of Europa is believed to be an ocean. The thick cover of ice over a liquid ocean may be a fertile site for life. The Antarctic subglacial lakes have similar basic boundary conditions to Europa.

An investigation of Antarctic subglacial environments should target the unique role these lakes may have in terms of the triggers for rapid evolutionary radiation, for understanding the global carbon cycle through major glaciations and as an analog for major planetary bodies...

After a short description of what we already know about Lake Vostok (see above) the authors continue with Part 2.

Part 2 (Bell and Karl, 1998, p. 5). Prolonged preservation of viable microorganisms may be prevalent in Antarctic ice-bound habitats. Consequently, it is possible that micro-organisms may be present in Lake Vostok and other Antarctic subglacial lakes. However, isolation from exogenous sources of carbon and solar energy, and the known or suspected extreme physical and geochemical characteristics, may have precluded the development of a functional ecosystem in Lake Vostok. In fact, subglacial lakes may be among the most oligotrophic (low nutrient and low standing stocks of viable organisms) habitats on Earth. Although "hotspots" of geothermal activity could provide local sources of energy and growth-favorable temperatures, in a manner that is analogous to environmental conditions surrounding deep sea hydrothermal vents, it is important to emphasize that without direct measurements, the possible presence of fossil or living micro-organisms in these habitats isolated from external input for nearly 500,000 years is speculation.

Lake Vostok may represent a unique region for detailed scientific investigation for the following reasons:

- It may be an active tectonic rift which would alter our understanding of the East Antarctic geologic terrains.
- It may contain a sedimentary record of Earth's climate, especially critical information about the initiation of Antarctic glaciation.
- It may be an undescribed extreme Earth habitat with unique geochemical characteristics.
- It may contain novel, previously undescribed, relic or fossil micro-organisms with unique adaptive strategies for life.
- It may be a useful Earth-based analog and technology "test-bed" to guide the design of unmanned, planetary missions to recently discovered ice-covered seas on the Jovian moon, Europa.

These diverse characteristics and potential opportunities have captivated the public and motivated an interdisciplinary group of scientists to begin planning a more comprehensive investigation of these unusual subglacial habitats. As part of this overall planning effort, a NSF-sponsored workshop was held in Washington, D.C. (7–8 November 1998) to evaluate whether Lake Vostok is a curiosity or a focal point for sustained, interdisciplinary scientific investigation. Because Lake Vostok is located in one of the most remote locations on Earth and is covered by a thick blanket of ice, study of the lake itself that includes *in situ* measurements and sample return would require a substantive investment in logistical support, and, hence financial resources.

Over a period of two days, a spirited debate was held on the relative merits of such an investment of intellectual and fiscal resources in the study of Lake Vostok. The major recommendations of this workshop were:

To broaden the scientific community knowledgeable of Lake Vostok by publicizing the scientific findings highlighted at this workshop... The goal of the workshop was to stimulate discussion within the U.S. science community on Lake Vostok, specifically addressing the question: "Is Lake Vostok a natural curiosity or an opportunity for uniquely posed interdisciplinary scientific programs?"

Part 3 (Bell and Karl, 1998, pp. 6–7). The discussion of the major obstacles to advancing a well developed scientific justification and plan to study Lake Vostok hinged on several major factors including:

- The exploratory nature of the program coupled with the paucity of data about this unknown region making development of a detailed scientific justification difficult; the need for technological developments to ensure contamination control and sample retrieval, recognizing that Lake Vostok is a unique system whose pristine nature must be preserved.
- The need for a strong consensus within the U.S. science community that Lake Vostok represents an important system to study, and recognition that international collaboration is a necessary component of any study.
- The recognition that the logistical impact of a Lake Vostok program will be

significant and that the scientific justification must compete solidly with other ongoing and emerging programs.
- That the lack of understanding of the present state of knowledge of the lake as a system within the U.S. science community remains a difficulty in building community support and momentum for such a large program.

These obstacles were addressed in workshop discussions and are specifically addressed in the report recommendations, the draft science plan and the proposed timeline.

Part 4 (Bell and Karl, 1998, p. 7). The overarching goal of the science plan is to understand the history and dynamics of Lake Vostok as the culmination of a unique suite of geological and glaciological factors. These factors may have produced an unusual ecological niche isolated from major external inputs. The system structure may be uniquely developed due to stratification of gas hydrates. Specific scientific targets to accomplish this goal include:

- Determine the geologic origin of Lake Vostok within the framework of an improved understanding of the East Antarctic continent as related to boundary conditions for a Lake Vostok ecosystem.
- Develop an improved understanding of the glaciological history of the lake including the flux of water, sediment, nutrients, and microbes into the Lake Vostok ecosystem.
- Characterize the structure of the lake's water column, to evaluate the possibility of density driven circulation associated with melting–freezing processes or geothermal heat, the potential presence of stratified gas hydrates, and the origin and cycling of organic carbon.
- Establish the structure and functional diversity of any Lake Vostok biota, an isolated ecosystem which may be an analog for planetary environments.
- Recover and identify extant microbial communities and a paleo-environmental record extending beyond the available ice core record by sampling the stratigraphic record of gas hydrates and sediments deposited within the Lake.
- Ensure the development of appropriate technologies to support the proposed experiments without contaminating the Lake.

The proposed timeline for all this research was (Bell and Karl, 1998, pp. 7–8):

1999–2000	Planning year
	Modeling studies
	Develop international collaboration
	SCAR Lake Vostok workshop
	Begin technology development
2000–2001	Site survey year 1
	Joint NSF/NASA Lake Vostok Proposals
	Airborne site survey
	Preliminary ground-based measurements
	Preliminary identification of observatory sites

2001–2002	Site identification and site survey year 2
	Ground-based site surveys
	Complete airborne survey if necessary
	Test/access/contamination control at a site on the Ross Ice Shelf
	Finalize selection of observatory sites
2002–2003	*In situ* measurement year
	Drill access hole for *in situ* measurements
	Attempt *in situ* detection systems to demonstrate presence of microbial life
	Install a long-term observatory
	Acquire microscale profiles within surface sediments
	Conduct interface surveys (ice–water and water–sediments)
	International planning workshop (including data exchange)
2003–2004	Sample retrieval year
	Acquire samples of basal ice
	Acquire samples of water and gas hydrates
	Acquire samples of surface sediments
	Stage logistics and second observatory
	International planning workshop (including data exchange)
2004–2005	Installation of second long-term observatory
	Analysis of data
	Building new models
	International planning workshop (including data exchange)
2005–2006	Core acquisition year
	Begin acquisition of long core
	International planning workshop (including data exchange)

At the time of writing it is the beginning of 2006. It is worth saying that none of the points in this 5-year timeline, have been undertaken. Chapter 11 explains partly why this is the case.

11

Penetration of Lake Vostok

Lake Vostok has not been penetrated yet. Soon we will obtain samples of the water from the lake, investigate the water depth via remote systems, and discover their implications. Real experimental data on salinity and currents in the water of the lake, and data on biological materials, if found, might change some understanding of the phenomena of this remote and unknown part of the Antarctic Ice Sheet.

Legitimate questions exist regarding the penetration of the lake, namely why, where, and how? We already know the size of the lake, salinity of its water (most likely freshwater), and some information on its currents. The study of accreted lake water ice indicates that a number of types of bacteria in the lake are similar to those in the overlying ice cores. It might be better to not penetrate the lake at all, or defer the event (Roura, 1999), but if it is decided to proceed, there should be procedures established not only for the physical penetration of the lake but also to ensure that penetration leads to as little harm or impact as possible.

Deep-core drilling at Vostok Station was terminated at a depth that left only 130 m of ice between the bottom of the hole and Lake Vostok, and a team of experienced scientists with special equipment is presently preparing to go farther. Other teams of international scientists are planning devices for penetration of the lake or to study samples that will be retrieved. The time of penetration and sampling is imminent, with the results having a major significance for many aspects of science. It is not important whether it is the British, Americans, Russians or scientists from other countries who will be the first to penetrate the ice sheet and probe the lake, however, concepts of penetration and exploration of the lake are currently different in Russia when compared with other countries.

The "Russian concept" is based on more than 30 years of experience in core drilling, including microbiologically clean core drilling of deep boreholes at Vostok Station, resulting in a vast amount of experience that is vital for balancing the characteristics of the drilling equipment designed to survive in rigorous Antarctic conditions, and the demands to minimize contamination of the lake. This balance will be achieved even if some contamination of the lake might inevitably occur. It should also be mentioned that micro-organisms, which might enter the lake due to possible contamination from a borehole liquid and drilling device (in spite of its cleaning) would have a very small chance of survival in a medium that is foreign to them, a medium that is devoid of nutrients.

It is also clear that the experience of the sole team that has been working for more than 30 years on the issues of drilling to 3,620 m, with the borehole remaining open, at the only working station (for more than 45 years) in the vicinity of Lake Vostok, is critical to continuation of the drilling of the last few meters of ice remaining above the lake. The team concept in this case, consisting of scientists and engineers who have worked on these issues for many years, dictates that the same individuals and their cumulative knowledge are vital for the continuation and completion of the work.

Another concept is based on the possibilities of the penetration of Lake Vostok from another location, and not from Vostok Station. The resulting time involved in this proposal would lead to years of logistic preparations related to construction of a new manned station in Antarctica, and related deep drilling. This could lead to potential contamination of the lake by virtue of the introduction of hundreds of tons of equipment, fuel, drilling liquid, food, and a number of individuals that would occupy the site for many months, to say nothing of the millions of dollars spent to achieve goals that might result in the uncontrollable contamination of Lake Vostok water. Goals that are questionable could be solved by using the existing Vostok Station as a base for the final drilling of ice and exploration of the lake. In favor of the latter, a rule of science states that if it is possible to conduct an "experiment" (in a philosophical understanding of the word), it should be done as soon as possible if there is a scientific demand for it. The above discussion favors this demand, as shown by the scientists of many specialties who are awaiting samples of water from Lake Vostok.

The strategy of penetration into Lake Vostok was presented by the President of the Scientific Committee on Antarctic Research (SCAR), Professor Rutford, before participants of the SCAR International Workshop on Subglacial Lake Exploration in Cambridge, England, September 1999. Professor Rutford reminded the audience of the experience gained from other projects, especially from the Ross Ice Shelf Project (RISP). The majority of participants might not have taken the comments seriously, but I did because I took part in this project from 1974–1978, when Professor Rutford was the Director of RISP. The major objective of the project was to drill through the 416 m thick ice shelf to penetrate into the sea from a location known as the "J-9" Camp, located 500 km south of the ice shelf barrier. There was no information at that time on whether life might exist beneath the ice shelf far from the open ocean. A group of anonymous experts decided that requirements for the drilling equipment for penetration of the sea below the ice shelf should include an avoidance of any contamination of the subglacial environment beneath the bottom of the Ross Ice Shelf. As a result of these unrealistic requirements made by experts (some from fields of science not related to the issue at hand), the Americans produced very expensive and sophisticated equipment that was incapable of working in actual Antarctic conditions. As a result, the drilling equipment became frozen in the borehole and became unworkable at the beginning of drilling operations in the summer 1977 field season.

Drilling of the access hole through the ice shelf was completed the next season on 2 December 1977, but this time with a flame-jet, produced by the continuous combustion of fuel oil in a high-pressure air flow. The 416 m deep access hole with a diameter of more than half a meter flame-drilled its way to the sea below in 7 hours (Browning, 1978). But it also filled the hole with a large amount of a mixture of fuel oil and soot from the incomplete combustion of fuel oil. As a result a large amount of it escaped into the sea beneath the ice shelf. No one seriously considered the issue of contamination, but Mr. Hansen, a famous drilling equipment engineer and designer of the drilling equipment that froze in the borehole while attempting to make an ecologically clean hole through the ice shelf, mentioned to everyone at the camp that "If they had allowed me to put gallons of contamination under the ice shelf, I would design a completely different and much more reliable drilling unit, which would have never failed!" However, nobody listened to him: "Do not spoil the victory, Hansen..."

THE FIRST ENTRY INTO THE LAKE FROM THE VOSTOK STATION BOREHOLE

Eager to drill deeper and end the 30 year long saga of drilling at Vostok Station by finally penetrating the lake, and following the recommendations of the Cambridge (1995, 1999) and St. Petersburg (1998) workshops to develop methods of entry into Lake Vostok and sampling to avoid altering the nature of the lake, the drilling teams of St. Petersburg Mining Institute, and the Arctic and Antarctic Research Institute (AARI) (Russia) proposed a project entitled: "Justification and development of ecologically clean technology for entry into the subglacial Lake Vostok".

The basic idea of this project included the use of the 3,623 m deep borehole 5G at Vostok Station, which had been drilled by teams of these two institutes. It was proposed that the last 130 m or so of ice should be penetrated using a special silicon–organic drilling liquid, chosen to allow a minimum of biological, physical, and chemical contamination of the lake.

The drilling device, used for electro-mechanical core drilling of the main part of the borehole, will have to be replaced with a new thermo-drill device that would melt the last tens of meters of the borehole to the ice–lake water interface without taking a core. The drilling procedure would keep the thickness of the silicon–organic liquid layer equal to about 100 meters, and the pressure within the borehole will be kept slightly lower than the water pressure at the ice–lake interface.

In this case, lake water will be pushed up by the difference in pressures in the lake and in the borehole when a thermo-drilling device penetrates the lake. The pressure difference will allow lake water to rise up about 50 m above the ice–lake interface. After that the thermo-drill device will be taken to the surface and the borehole, with a 50 m thick layer of lake water and about a 100 m thick layer of silicon–organic liquid above it, will be left until the lake water within the borehole is completely frozen. A core-drilling device will then be lowered to the bottom of the borehole and some tens of meters of the frozen lake water will be taken to the surface of the ice sheet in the form of an ice core.

This part of the project is worth looking at more closely because of its importance (Verculich *et al.*, 2002). Three steps would be made to enter the lake.

(1) Core drilling of the upper 100 m of remaining ice will be performed by the same electro-mechanical drill that was used for the previous core drilling. It was planned that the first 50 m of ice would be drilled in the field season of 2004–2005 and the next 50 m in 2005–2006. However, due to difficulties in getting new supplies to Vostok Station the drilling of the first 50 m of ice was postponed to 2005–2006.

Precious samples of accreted ice will be divided into three equal parts and sent to France, the U.S.A., and Russia for study. The results of this study will be used for possible amendments to the entry procedure.

(2) A specially designed container will transport an ecologically clean, hydrophobic liquid to the bottom of the hole, possibly a silicon–organic oil with a density lower than that of the lake water, but higher than that of drilling fluid. This liquid will provide a "buffer layer" about 100 m thick at the bottom of the hole (Figure 11.1). The pressure at the bottom of the hole (P_{hole}) will be kept lower by about 0.3 Mpa than lake water pressure at the ice–water interface (P_{lake}) below Vostok Station.

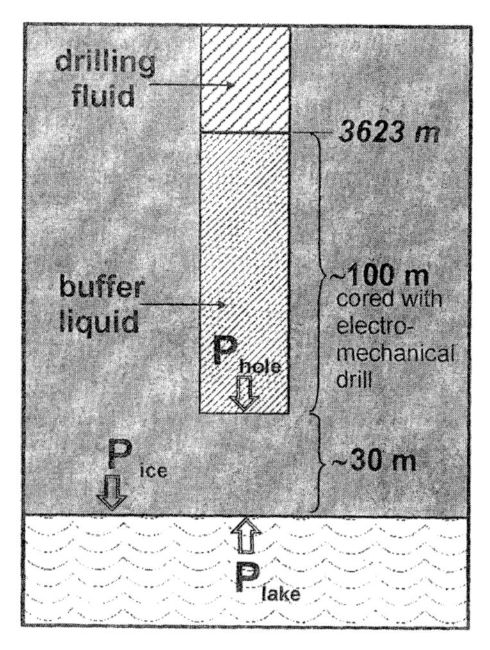

Figure 11.1. Diagram of the first step of the approach to enter Lake Vostok from the borehole at Vostok Station.

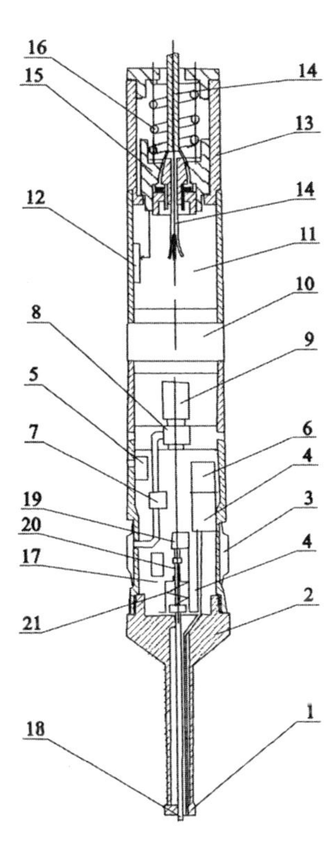

Figure 11.2. Thermo-electrical drilling device for the initial entry into Lake Vostok (after Verculich *et al.*, 2002). The drill will melt the last 30 m of the ice in non-stop mode with a speed of about 4 m hr^{-1} until reaching the ice–water interface. Its front (lowest) pilot-heated bit (1) will enter the lake. The device is 7 m long and its diameter is 0.132 m. A description of the drill's component parts is given in the text.

(3) A specially designed and ecologically clean thermo-electrical drilling device will drill the last 30 m of ice to the ice–water interface and enter the lake. This device will be about 7 m long and its diameter will be 0.132 m (Figure 11.2).

This drill will melt the ice by means of a small pilot-light-heated bit 50 mm in diameter (1) and a main heated drilling ring 132 mm in diameter (2) with a truncated cone form. The pilot-heated bit will be located 2 m ahead of the main heated drilling ring. Packer (3), pressure sensors (4, 5, 6), valve (7), pump (8) driven by motor (9), electronic package (10), electrical compartment (11), hole bottom load sensor (12), cable lock (13) with the cable (14), moveable bushing (15), and spring (16) will be located in a cylindrical container behind the main heated ring drill. The contact sensor (17) consists of a stock (18), sensing elements (19, 20), and a spring (21). This spring will keep the stock (18) in such a position that its end will be ahead of the pilot-heated drill bit when it will penetrate the ice.

This device will probably penetrate the last 30 m of ice in the Antarctic

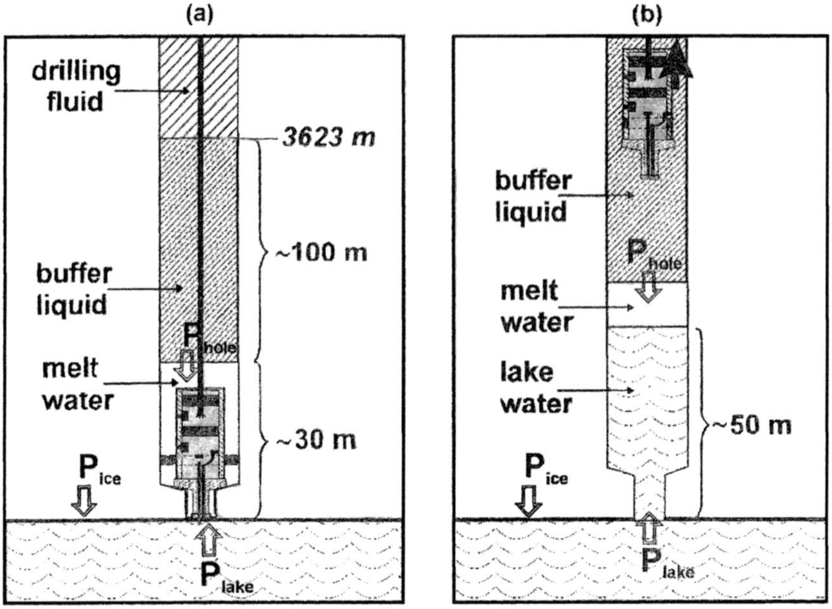

Figure 11.3. When the tip of the pilot-heated bit of the device penetrates the ice sheet and enters the lake, the stock ((18) in Figure 11.2) loses support from the ice underneath and is pushed down by the spring, tripping the sensor elements and turning on the contact sensor. In response the pump's motor starts, the pump starts drawing water into the packer, which closes contact between the lake water below the packer and water above it, electrical heating is terminated, and movement of the device downward stops (a). A proper pressure difference (P_{hole} less than P_{lake} by appropriate magnitude) will be checked, the packer will be removed, the drilling device will be raised and the lake water will enter the hole, filling about 50 m of it (b). The lake water will then freeze within the hole (after Verculich *et al.*, 2002).

summer of 2007–2008 in non-stop mode at a speed of about 4 m hr^{-1}. The hole melted by this device will consist of two parts: (1) the lower part, with a length 2 m and diameter 50 mm, and (2) the upper part, the main part with diameter 132 mm. The stock (18) of the pilot-heated bit will be pushed against the ice at the bottom of the hole to its lower position while the device proceeds downward through the ice. When the tip of the pilot-heated bit penetrates the ice sheet and enters the Lake, the stock will lose ice support and will be pushed down by the spring (21), tripping the sensor elements and turning on the contact sensor (17). In response to this the pump's motor will start, drawing water into the packer, which closes the contact of lake water below with the water above it, then the electrical heating will be terminated, stopping downward movement of the device (Figure 11.3(a)).

Data from pressure sensors will be obtained and analyzed at the surface of the ice sheet and if pressures of the hole (P_{hole}) and of the lake (P_{lake}) are different from those originally estimated, the drilling liquid level will be changed to get the proper pressure difference (P_{hole} being less than P_{lake} by an appropriate magnitude). The packer will then be removed, the drilling device will be raised and lake water will rise into the hole, filling about 50 m of it (Figure 11.3(b)).

Designers of the procedure estimate that the lake water will enter the hole very slowly, because the pressure of the lake (P_{lake}) will be only slightly higher than the hole (P_{hole}) at the beginning of this process and will decrease to zero at the end. Lake water while filling the hole has to move through a narrow space between the drilling device and the walls of the hole and then raise 50,000 kg of drilling fluid within the hole above it.

Calculations and numerical modeling show that the lake water entering the hole will become frozen in the hole within a day.

(4) An electro-mechanical core-drilling device will then be inserted into the hole, and an ice core of newly frozen Lake Vostok water will be taken and brought to the surface. The drilling procedure will be stopped above the bottom of the ice sheet leaving 10–15 m of ice between the bottom of the hole and the Lake Vostok water beneath.

The device for entering the lake is presently being manufactured, although the designers are already working on improvements to the original concept (Verculich *et al.*, 2002). These include further development of procedures for cleaning and sterilization of the device, a search for a better "buffer liquid", plus theoretical and experimental checks of the possibility for complete collapse of the 2 m thick ice structure below the main ring drill bit during the last phase of thermal drilling. Other improvements or changes include the possibility of upgrading the device to first take a clean sample of the lake water *in situ*, directly from the lake. This option is important because it would allow comparison between the *in situ* sample with that of the lake water ice core which would be extracted later in the process.

The designers of the equipment also consider it necessary that the device and the actual procedure of penetration should be verified by laboratory tests and in the field on ice shelves or small subglacial lakes.

The Ministry of Industry, Science, and Technology of the Russian Federation (Russia) funded this project in 1999–2001, and an Expert's Commission in the course of the National State Ecological Expert Examination in March 2001 approved it. Dr. Abizov, the microbiologist who was the first to examine Vostok ice cores biologically to 3,000 m, Professor Kudryashov, a leader of the team of drillers at Vostok Station and designer and manufacturer of ice-drilling devices for more than 20 years, academician Kotlyakov, who for many years was the Russian Permanent Representative for SCAR, Dr. Lukin, the Head of the Russian Antarctic Expedition (RAE), the author of this book, plus other scientists experienced in problems of environmentally clean ice drilling met with the experts and participated in the State Examination, reaching a consensus for the approval of the project.

Some experts advise that the next stage in advancing this project could be its testing at sites other than Lake Vostok, where full environmental monitoring of the procedures would be possible. One of the sites suggested is the ice shelf near the Russian Novolazarevskaia Station, a possible proving ground for field tests. So far this has not occurred because of financial problems. The author's personal experiencc suggests that this testing would be useless because there is no way to check experimentally the degree of possible contamination of subglacial water as the drilling device penetrates into the water.

The project itself and the "expert conclusions" were presented to the participants at the XXIV Antarctic Treaty Consultative Meeting, held in St. Petersburg in 2001, attracting the attention of the international scientific community.

THE FIRST STEP TOWARD PENETRATION

The first step toward accomplishing this project should be core drilling of the first 50 m below the bottom of the existing borehole. In this case further drilling of 50 m of accreted ice will reduce the remaining ice thickness to 80 m, which still exceeds, by a factor of 3, the minimum allowed thickness (25 m) established and agreed at the Lake Vostok workshop held in Cambridge in 1995. With this in mind, the RAE planned to do this drilling in hole 5G from 3,623 m depth to about 3,673 m depth as fast as possible. Many specialist scientists have requested this segment of the core for study. This part of the project leaves sufficient ice below the bottom of borehole 5G to perform this aspect.

This plan was presented by Dr. Lukin, Head of the RAE, at the first meeting of the Subglacial Antarctic Lake Exploration Group of Specialists (SALEGOS), held in Bologna, Italy, in November 2001 with the hope of quick approval. SALEGOS was formed by SCAR in 2000 on a recommendation of the Workshop on Subglacial Lakes held in 1999 in Cambridge to deal with everything connected with Lake Vostok. Dr. Lukin proposed the plan to retrieve an additional 50 m of suspected accreted lake ice from the Vostok borehole in the 2002–2003 field season as part of a general project to enter the lake (Kennicut, 2001).

The plan was supported by the results of a glaciological study described by Lipenkov (private commun., 2001):

> *The vertical seismic profiling experiments (Masolov et al., 2001) and the borehole temperature measurements (Salamatin, 2000) performed repeatedly in the hole 5G-1 to the bottom at 3,623 m suggest that the thickness of the remaining ice that separates the hole from sub-ice water is 130–140 m, corresponding to a total thickness of the ice sheet of 3,750–3,760 m. From ground-based radar sounding (Popov et al., 2000), the ice sheet thickness in the vicinity of the hole is 3,775 ± 15 m, corresponding to 140–170 m of remaining ice.*
>
> *The lower layer of the accreted ice is characterized by exceptionally large ice crystals (from 10–100 cm and larger). The high crystalline quality, as revealed by recent X-ray diffraction measurements of these crystals (Montagnat* et al.*, 2001), indicates that lake ice is not plastically deforming*

under in situ conditions. This result supports earlier assumptions that accreted ice in the vicinity of Vostok Station is blocked by the lake edge (Lipenkov and Barkov, 1998). In addition, low lattice distortion (low dislocation density) of these crystals rules out significant diffusion of the drilling fluid through the ice lattice.

Given that the lake ice is formed by slow freezing of a mixture of frazil ice plus host water (Souchez et al., 2000), one can expect polycrystalline ice with crystal sizes not exceeding some millimeters immediately after freezing. Crystal growth should therefore occur after initial freezing and during the long annealing time while lake ice remains at a temperature close to the melting point. It is considered that the abnormal grain growth is probably at the origin of the structure of lake ice with exceptionally large and high crystalline quality grains typically characterized by well ordered grain boundaries (Montagnat et al., 2001).

Both well ordered (low-energy) grain boundaries and very large crystal size (extremely low vein density factor) make the permeability of accreted ice to drilling fluid very unlikely. We should therefore not fear for the long-term contamination of sub-ice water due to drilling fluid intrusion from the borehole bottom, at more than 80 m above the lake surface.

After reviewing and discussing background material for the proposal of the "Russian plan to deepen the Vostok hole", "...the group agreed that the scientific justification for further drilling of 50 m of accreted ice was adequate to endorse this deepening to get additional accreted lake ice for coordinated international studies. However, the group did not feel that the necessary glaciological experts were present for a valid technical and environmental judgment to be made, and that an independent assessment of the drilling plan was needed in order to judge the safety of the proposed deepening of the Vostok borehole. The convener will correspond with geophysical and glaciological experts on this issue and request their input on the safety aspects of the plan for deepening the borehole. If at all possible this issue will be resolved intersessionally, and if not this item will be placed on the agenda for further consideration for the proposed spring 2002 meeting of the group" (Kennicut, 2001).

What happened next is outlined in the following letter from the convener of the group, John Priscu, Professor of Ecology at Montana State University (U.S.A.), to the President of SCAR, Dr. Rutford in April 2002, about a month and a half before the next meeting of SALEGOS.

The RAE presented plans at our last SALEGOS meeting to retrieve an additional 50 m of accretion ice from the deep Vostok borehole during the 2002–2003 season. Details of this plan are outlined in the SALEGOS report from meeting I (Bologna, Italy, 29–30 November 2001). The group agreed that the scientific justification was adequate to endorse a deepening of the Vostok borehole to provide additional accreted lake ice for coordinated international studies. However, the group did not feel that appropriate experts were present for a valid technical and environmental judgment to be made. I, as convener, was tasked with obtaining statements from geophysical and glaciological experts concerning environmental issues. I presented five experts (only four responded) with the Russian plan to deepen the Vostok borehole by another 50 meters. The letter sent to each expert read as follows.

Dear Colleague,
You may know that the Russian program is planning on taking another 50 m of accretion ice from borehole 5G (leaving 80 m between the bottom of the borehole and the ice–water interface). SCAR has requested that SALEGOS provide recommendations concerning the scientific and environmental issues that surround such a venture. There was a unanimous consensus that more accretion ice, particularly the very clear ice below 3,608 m, will have important scientific ramifications. However, the environmental issues remain unresolved. I was tasked by the Group of Specialists to obtain the judgment of experts on this issue. In this respect, would you please send me your formal evaluation of the environmental issues that you feel may result from increasing the depth of the 5G borehole? Specific items you should address (but are not limited to) are:

(1) What is the probability of drilling fluid permeating the additional 80 m, thus contaminating the lake water?
(2) What is the probability that the same mechanical properties of the accretion ice previously collected (the clear ice between 3,608 and 3,623 m) exist to the lake water interface?

The response from each individual is attached (because several experts requested to remain anonymous I imposed anonymity on all). Based on the advice and comment from these recognized specialists and discussions among the Group of Specialists, we are unable to arrive at a consensus on the risk of contaminating the lake during a deepening of the existing hole. This is primarily due to a lack of adequate information about the properties of the ice to be drilled and an inability to quantitatively predict ice behavior at these great depths and unusual conditions. One expert suggests that based on experience with drilling-induced hydrofracturing that any fracture once initiated would propagate long distances. Grain boundary percolation of contaminants also raises concerns about lake contamination, as the expectation is that increasingly warm temperatures will be encountered as the lake surface is approached. Other experts propose various preventive activities that might reduce the risk of lake contamination. While recognizing that the Russians are world leaders in ice drilling, the general consensus of other experts is that additional assessment would be needed before the risk of lake contamination can be adequately determined. The precautionary approach would seem prudent but we defer to SCAR to determine what the next steps might be. I suggest that the SCAR Executive convene a meeting in Shanghai to discuss these issues in detail and develop clear and tractable guidelines that can be presented to the Russian program.

The proposed drilling at Vostok Station to deepen the borehole by 50 m in 2002–2003 was thus postponed until 2003–2004 due to the absence of a clear answer as to whether the borehole can be deepened by another 50 m, because the anonymous experts were uncertain about the potential for contamination of the lake. The XXVIth Consultative Meeting of the Antarctic Treaty agreed in 2003 with the "Russian proposal". "It was like a competition to be first to the Moon. We have abided by the rules of the game within the international Antarctic Treaty", said Dr. Lukin, Head of the RAE, to the Russian News Agency "RIA Novosti" on 13 May 2004.

Dr. Lukin advised that the first 50 m of the Vostok borehole would be drilled in 2004–2005, leaving about 80 m of the ice. In 2005–2006, another 50 m of ice would be drilled and "The lake waters will be reached in 2006 or 2007" ("RIA Novosti", 13 May 2004).

Now, at the end of January 2006, we know that this plan was partly completed. A new stage of drilling at Vostok Station began in December 2005 and ended in the middle of January 2006 with 27.5 m of the core taken to the surface. Each round of drilling took about 6 hours and brought to the surface about 1.5 m of the core, increasing the depth of the hole from 3,623 m to 3,650 m. This does not represent a big step, but nonetheless a very important one. Part of this core was sent to Europe for study.

WHEN THE LAKE IS PENETRATED

The most significant part of this scientific story is forthcoming. There will soon be samples of water from the lake, and associated investigations of the water thickness via remote systems to determine any implications. Experimental data on salinity and currents in the water of the lake and data on biological materials, if they are found, will possibly change some understandings of the phenomena.

In situ observations of the water of Lake Vostok

Initially, a static string approach would be used consisting of instruments placed in the hole in order to get a time series of measurements at different depths and sides of the lake (e.g., at the south end (beneath Vostok Station), where freezing of the ice roof of the lake occurs and at the opposite side of the lake, where there is melting of the ice roof). Dissolved oxygen and other important gases, conductivity, pressure, temperature, water speeds, and turbidity would be measured in time intervals to gather information on water circulation and the geophysical, geochemical, and biological regimes of the lake. At the bottom of the vertical string of instruments, a redox electronic system will be installed for measurements of surface sediments and various biogeochemical processes. Penetrometers or shear vane devices will also be installed to measure sediment compaction (Kennicut, 2001).

Water sample return

A small volume of water will be transported to the surface of the ice sheet at first. *In situ* filtering will be completed to recover microbial biomasses for DNA and other analyses. A 400-bar pressure change during the samples return to the surface is not considered to be a problem for microbial studies, but temperature changes from the freezing point of water at the ice–lake water interchange (2–3°C) to the ice sheet surface (approx. −55°C) will be a problem due to freezing of samples or the possible exposure of samples to high temperatures in the case of hot-water drilling. Samples will probably need insulation to avoid such problems. DNA analyses, culturing of microbes, and a full spectrum of microscopy should be used for sample analyses. Molecular probing might be employed to identify specific microbial groups. Large pressure changes while taking water samples from the lake to the laboratory may influence gas analyses requiring special methods of sample retrieval to be developed. A wide spectrum of organic and inorganic chemical analyses of filtered and unfiltered samples will be performed (Kennicut, 2001).

Sediment sample return

Sediment–water interface samples and shallow-depth sediment cores will be taken for information on biological and biochemical conditions, processes, and cycling. Longer cores from deeper horizons will be taken to establish the history of the lake and for paleoclimate studies. If coring is done from the surface, a cased coring system will be used to decrease contamination due to redistribution of materials within the lake. Temperature and pressure changes as a result of raising the cores from the bottom of the lake to the ice sheet surface also have to be taken into account. Analyses of DNA, culture microbes, metabolic activity measures, and a full suite of microscopy and chemical analyses will be completed on the cores. Experience gained from the Ocean Drilling Project, as well as retrieving sea bottom sediment cores through the 415 m thick ice of the Ross Ice Shelf is valuable for application to the Lake Vostok project (Kennicut, 2001, 2002a,b, 2003a,b).

The group of specialists determined the following priorities:

- *To determine the identity and diversity of life forms.* Answers to the major question of whether the biology of Lake Vostok is viable or fossil can be found here. This study includes determination of the amount of biomass and density of each type of life; determination of progenitors of subglacial life in the lake water, including hydrothermal sites, ice, hydrate crystals, determination of the types of organisms that are metabolically active, and determination of the spatial location of diversity.
- *To define the redox couples that support life.* Included here is the elucidation of any unique biochemical or physiological processes; determination of *in situ*

growth and metabolic rates of organisms; determination of the minimum amount of energy required for growth; determination of energy sources and how energy is extracted from the environment; determine the carbon sources that support life in the lake; and investigation of connections between living organisms in the lake and gas hydrates.
- *To determine the evolutionary history of Lake Vostok and its biota through the study of bottom sediment cores.*

All the above need to be done with critical testing, verification, and monitoring for potential possibilities of contamination during the drilling and penetration of the lake, work within the lake, and return of equipment, samples, and cores from the lake to the surface through about 4,000 m of ice sheet. Stewardship issues include the protection of Lake Vostok environments by ensuring minimal alteration or change due to scientific studies. From a scientific point of view it is important that samples be available for study in pristine condition and remain unchanged from their sites of sampling, and that the presence of manufactured devices does not influence the results of measurements and observations. The present debate and controversy on the origin of microbes in the accreted lake ice are examples of the potential problems that must be dealt with during the lake's exploration. Another area of concern is that previously unknown microbes and other microbiological material be properly secured and stored to avoid unwanted contact with our environment. Redistribution of lake water and sediments during *in situ* measurements and observations should be kept to a minimum because of the oligotrophic nature of the lake (Kennicut, 2001).

APPLICATION OF EXPERIENCE GAINED FROM PLANETARY PROTECTION

There is a long history of planetary protection used by space programs. The SALEGOS group of specialists discussed the possibility of using experience gained from such protection for developing a framework for constructing a procedure for the defense of Lake Vostok against contamination. It is considered that the challenges faced by subglacial lake exploration resemble those of planetary exploration. The essential goal of planetary protection is to prevent contamination by foreign organisms during exploration for life – the goal in subglacial lake exploration is the same. The U.S. National Research Council program entitled "Preventing the Forward Contamination of Europa" may be used as an example because their procedures are beneficial toward understanding the applicability to Lake Vostok.

NASA recommends that components of the equipment should be cleaned often and individually using different techniques depending on the durability of the component. Current Mars mission protection requires that missions not carrying life detection experiments must be cleaned to ensure that the total "bioload" does not exceed 300,000 spores and the "density" of the spores on the surface does not exceed 300 spores per square meter. Missions comprised of life detection experiments should be put through additional cleaning to ensure that the total "bioload" does not exceed 30 spores. Efficiency of these cleaning operations depends on the types and numbers of micro-organisms, their resistance to treatment, and comparability between the device being cleaned and technique being used. In our example there is a conclusion that for each future mission to Europa the probability of contamination of a subglacial ocean of Europa by viable micro-organisms brought from the Earth should be less than one in ten thousand. Problems of biological contamination connected with Europa are discussed in detail by Richard Greenberg in his book *Europa, the Ocean Moon.* The group of specialists on planetary studies suggested a procedure for a calculation to determine whether each particular Europa mission met the mission's requirements and a group of specialists on the study of subglacial lakes (SALEGOS) should do the same.

The group was unable to reach complete agreement on cleanliness standards for "bioloads" for subglacial lake studies. The agreed basic concept is that "... the 'bioload' on any instrument package should be at such a level that incidental transport of organisms into the study site is reduced to a sufficiently low probability" (*Kennicut*, 2001). However, this declaration means nothing, because we still have no idea about specifying "such a level" in quantitative form. The author's point of view is closer to the view of the minority of the SALEGOS group, which thinks that organisms, which can be introduced into the Lake Vostok environment by contamination due to penetration into the lake, have a very low probability of surviving and multiplying in the lake's truly alien environment (Kennicut, 2001).

Other projects planning penetration into Lake Vostok

Up to now there is no other project planned for penetration into Lake Vostok that is sufficiently developed from a contamination point of view.

There are projects planned for the penetration of the Lake from some other locations on the ice sheet. The Lake Vostok workshop in Washington D.C. in 1998 began with a question: "Lake Vostok: A Curiosity or a Focus for Interdisciplinary Study" and ended with the timetable of events, connected with observatory sites, which should be erected at the surface of the ice above Lake Vostok over a 6-year period beginning in 2000 (Bell and Karl, 1998). The timeline proposed was:

- *2000 (2000–2001 field season).* Year 1 for preliminary identification of observatory sites.
- *2001 (2001–2002).* Year 2 for site identification and site survey, to include ground-based site surveys and testing of access/contamination control technologies at a site somewhere on the Ross Ice Shelf.
- *2002 (2002–2003). In situ* measurement year. A long-term observatory will be installed at a chosen site above the lake and an access hole will be drilled into the lake for *in situ* measurements. Attempts for *in situ* detection of microbial life, vertical profiling of the water characteristics, micro-scale profiles within the

bottom surface sediments, and ice–water and water–sediment interface surveys will be undertaken. An international planning workshop and data exchange will be held.

- *2003 (2003–2004)*. Vostok Lake sample return year. Retrieval of basal ice samples, samples of water and gas hydrates, and samples of lake bottom sediments will be undertaken. A search for a second observatory site will be performed. An international planning workshop and data exchange will be held.
- *2004 (2004–2005)*. The year of installation of the second long-term observatory. "Analyses of data. Build new models" – inform authors of the timeline for this year.

Presently (at the begining of 2006) not one of the timeline bullet points has been realized. Why is the gap between reality and the timeline so large?

No one from the RAE or from the cadre of Russian scientists involved in many years of study of Lake Vostok were present at the 1998 workshop on "Vostok Lake: Curiosity or a Focus of Interdisciplinary Study" and not a single word about the possibility of using Vostok Station's deep hole for international interdisciplinary study of the lake was published in its final report (Bell and Karl, 1998). This discrepancy occurred because, for the first time in the history of multi-international scientific exploration of Antarctica, scientists from different countries did not move toward assisting each other to proceed expeditiously, but instead were motivated by their own agenda to retard progress in order to achieve a time advantage toward being the first to penetrate the lake. This explains why the SALEGOS group of specialists failed to authorize permission (contamination clearance) from SCAR to drill an additional 50 m at Vostok Station. The clearance was received from SCAR after 4 years of discussions, amounting to the time allocated for the main part of the timeline program.

Deep ice core drilling of the kind performed at Vostok Station would not be a feasible option at a newly planned observatory. It takes years of drilling plus years of preliminary preparation. One available technique is deep drilling (actual penetration) of the ice sheet using hot water. This concept is as old as thermal drilling of the ice by a solid hot point or an electrically heated ring. Technical difficulties connected with this method, however, presented major problems, so that for a long time hot water drilling was used only for making small-diameter shallow holes.

The technology changed in 1978, when J. Browning, an engineer and inventor from Hanover, New Hampshire, U.S.A., applied the technique of flame-drilling through the Ross Ice Shelf at the site known as J-9. The equipment included a large, industrial-size, full output 250-horsepower water boiler heated by burning 300 kg of diesel oil per hour (Browning, 1978). This heated about 280 kg of water per minute from about 2°C to plus 98°C. Hoses brought the hot water to the bottom of the hole in the ice and made a hole with a diameter of 0.90 meters at a speed of about 40 m h^{-1}. Less than a day was required to produce a hole through the 416 m thick ice shelf and enter the sea below.

This technique, used to produce large, deep holes, was also developed for drilling more than 1,000 m to the bottom of the ice sheet of West Antarctica to reach the bottom of ice streams, as well as for the approximately 1,000 m hole at the U.S. Amundsen–Scott South Pole Station for use in the Antarctic Muon and Neutrino Detector Array (AMANDA) project. More than 10 deep holes, each more than 1,000 m deep and about 1 m diameter were drilled (Koci, 1999).

As a result of these successes, it appeared that rapid hot water penetration to the bottom of the 3,700–3,500 m thick ice cover of Lake Vostok could be achieved easily using this kind of drill, considered applicable for operating at the planned observatory sites.

However, a study performed recently by experts on hot water deep drilling through the Antarctic Ice Sheet showed that a hot water drill can provide rapid access holes to depths of 1.5 km. However, if the hot water drill is scaled to drill deeper, the energy requirements are such that they become enormous if depths below 1.5 km are desired (Clow and Koci, 2002). The same specialists showed that the new type of hot water drills required for penetration of Lake Vostok's ice cover would weigh about 200 tons and require about 50 tons of fuel, an unacceptable amount. The drilling itself would take only 10–12 days, but the hole of about 0.3 m diameter could only be used for study purposes for 2–3 days before it would close due to the surrounding pressure.

(Clow and Koci, 2002) recommend that only the use of so-called coil tubing drilling technologies (Gaddy, 2000), developed recently by commercial firms in the U.S.A. can be used for this kind of task (Figure 11.4).

The principle of this technology is the use of bendable, advanced composite spoolable tubing (Fowler *et al.*, 1999) that is capable of transporting a large amount of liquid, pumped into the hole by a high-pressure pump located at the surface. There is no drill housing or mast, as such, but an "injector" is used instead. The injector provides control for the drill string of bendable tubing. A fluid passing through the tube under pressure drives a downhole hydraulic "mud motor", which drives the drilling head. Core barrels, or special kinds of instruments can be attached to the drill head. The chips and fluid return to the surface outside the bendable tube using the annular space between the tube and the hole wall. The hole should be sealed through the firn layer to prevent drilling fluid leaking into the permeable firn. The chips are separated from the drilling fluid at the surface, and the drilling fluid is then pumped back down the hole (Fowler *et al.*, 1999).

The use of this type of technology for deep and rapid Antarctic ice drilling will need considerable field testing and additional studies. This technology was used in the Arctic for temperatures as low as −40°C at the surface and −15°C within the hole. The mean annual surface temperature of the ice overlying Lake Vostok is −56°C, being about the same for the upper 500 m of the hole, so it is impossible to predict the consequences of these temperature differences. Nevertheless, it might still be possible that advanced composite spoolable tubing technology of deep ice drilling may provide the required holes for new observatories above Lake Vostok, as suggested by the timeline of the "Curiosity or a Focus for Study?" workshop.

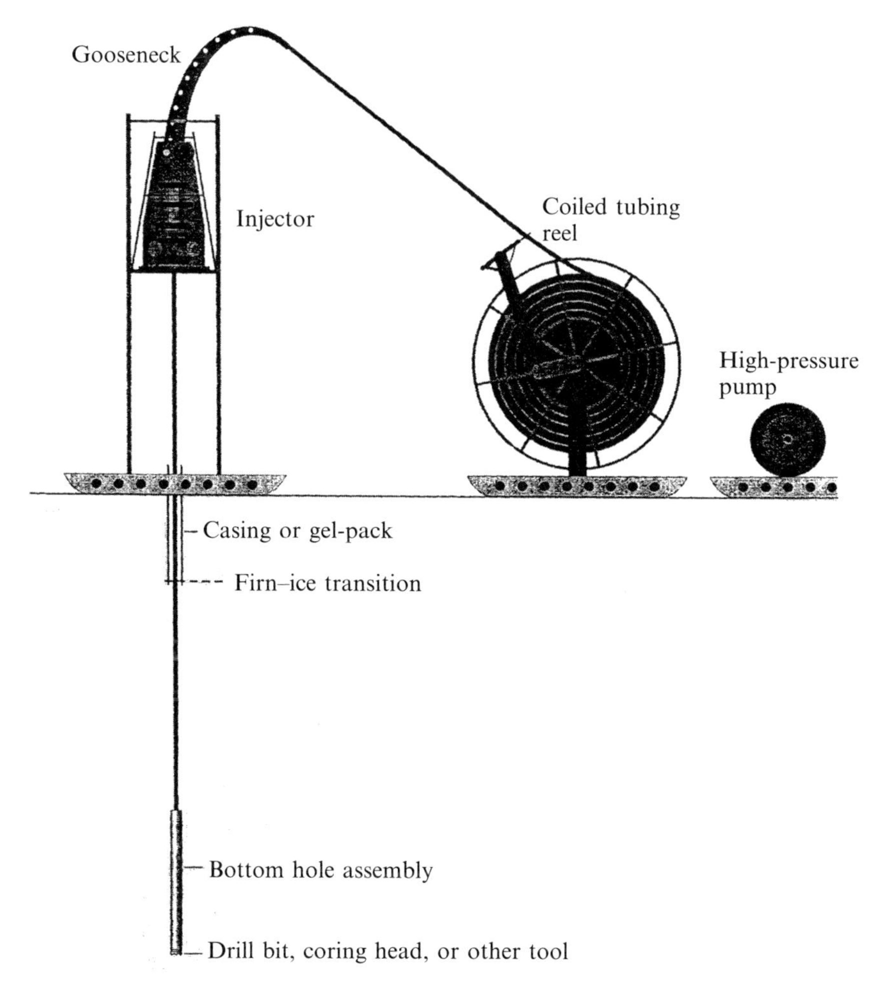

Figure 11.4. Coil tubing drilling technology proposed for fast access to Lake Vostok (after Clow and Koci, 2002). The main idea of this technology is the use of bendable, advanced composite spoolable tubing that is capable of transporting a large amount of liquid pumped into the hole by a high-pressure pump located at the surface.

THE USE OF VOSTOK STATION'S FACILITIES AS A FIRST STEP TOWARD THE INTERNATIONAL STUDY OF THE LAKE

Observation sites for additional deep drilling and penetration into Lake Vostok will probably not occur in the predictable future. It is illogical to spend money, time, and energy creating new ecologically unacceptable research stations (observatory sites) at the surface of Lake Vostok when there is a station already there (Vostok Station), as well as a stable hole nearly reaching the ice–lake boundary (Figure 11.5).

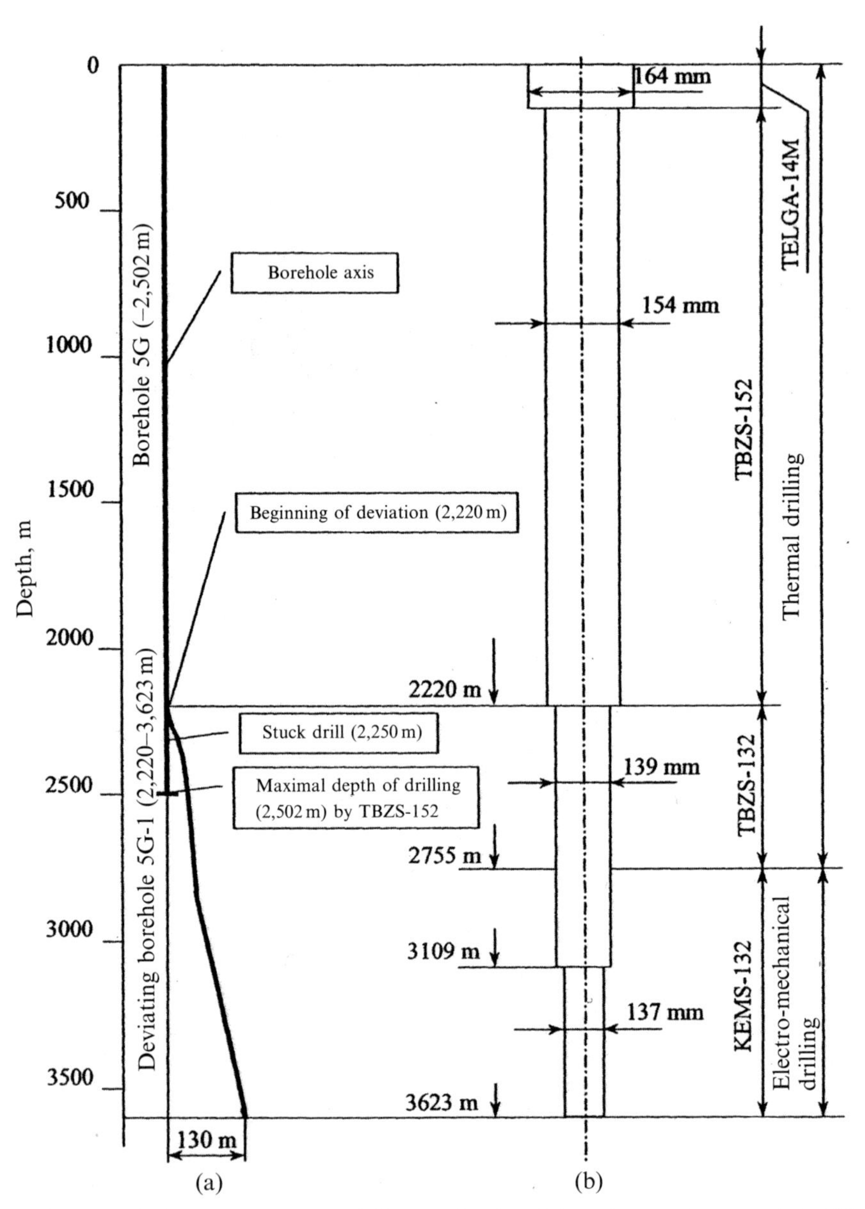

Figure 11.5. Boreholes 5G and 5G-1 of Vostok Station (after Kudryashov *et al.*, 2000): (a) inclination; (b) scheme of borehole.

This hole (5G) is the result of a complicated and careful drilling operation over many years. It was drilled at first with a 152 m diameter thermal drill to a depth of 2,500 m, producing a hole diameter of 154 mm. The drill became stuck in the hole at 2,250 m while raising it to the surface. The cable was severed at the top of the drill and removed from the hole. A deviation from the original hole 5G at approximately

2,200 m depth (50 m above the lost drill) was made and hole 5G-1 was drilled with another 132-mm thermal drill to 2,755 m. The diameter was enlarged to 139 mm. Ice coring of deeper horizons was continued with an electro-mechanical drill to today's depth of 3,623 m (Kudryashov *et al.*, 2000). The hole was filled with about 60 tons of drilling fluid to compensate for external pressure decreasing the diameter of the hole. For this purpose the density of aviation kerosene (the main component of drilling fluid) was increased by adding a densifier (Forane F-141b), resulting in a density of the fluid within the hole of 928 kg m^{-3}. The level of the liquid in the hole is maintained at a depth of 95 m from the surface. In this case overburden pressure of the ice is slightly higher (0.1 Mpa) than the hydrostatic pressure of the liquid at the bottom of the hole, therefore, any connection between the lake water and the liquid in the hole will not contaminate the lake, and hole closure at the bottom will be small, according to calculations – less then 0.1 mm yr^{-1}. This means that the hole will be open and ready for operations for any national and international projects for an indefinite period of time. The advantages provided by the existence of Vostok Station and an existing deep borehole should be considered before money is spent on the design of new stations (observation sites) and boreholes. To achieve this level of information and co-operation, all scientists must learn to work together again in a similar fashion adopted for the International Geophysical Year (IGY) of 1957–1958 and after. At that time, scientists of different nations working in Antarctica showed that combined team work was beneficial for all, and thus provided grounds for the Antarctic Treaty, despite the fact that two nations deeply involved in the study of Antarctica were ideologically divided by the Cold War (the U.S.A. and the Soviet Union).

The Antarctic Treaty and its spirit served as an example for successful international cooperation that has extended to space exploration and succeeding treaties on space. However, regarding scientific studies of Lake Vostok and penetration of the lake, the earlier spirit of the Antarctic Treaty has partly disappeared.

In a year a new scientific adventure will occur, the International Polar Year (IPY), which should provide the stimulus for a new generation of scientists willing to work cooperatively in Antarctica to study Lake Vostok.

A related project is one that was developed by the Jet Propulsion Laboratory (NASA) for penetration into the subglacial sea of Europa, an ice-covered satellite of Jupiter (Chapter 1) (Carsey and Horvath, 1996). It was discussed at a JPL sponsored Earth–Europa workshop that was held at JPL on 28–29 October 1996. The workshop was devoted to the possibilities of comparable studies of some features of both Earth and Europa that have distinct similarities. *Nature* announced the discovery of Lake Vostok just 4 months prior to the workshop, and interesting similarities between conditions in Lake Vostok and a "subglacial" sea below the newly discovered ice crust of Europa were the center of attention.

It was planned that devices, named "Cryobots" (Figure 11.6), would go from the ice surface of Europa to its "subglacial" sea, and from the surface of the Antarctic Ice Sheet to subglacial Lake Vostok, by melting–sublimating a cavern in the ice body and sinking its way to the water space below under gravity without maintaining a borehole in the ice behind them.

JPL *in situ* exploration
Antarctica as an analog for Europa

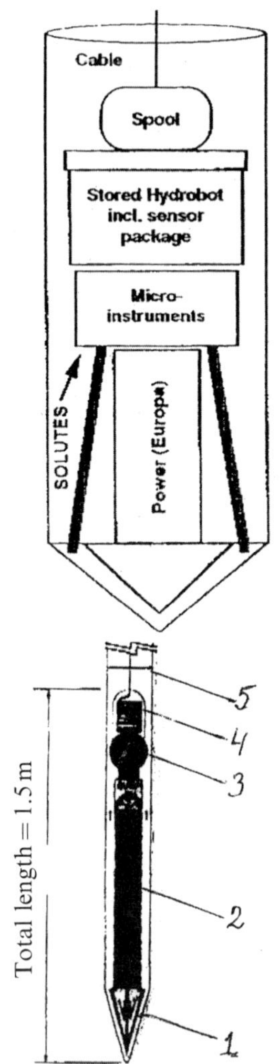

Science objectives under the ice

- Understanding possible parallel biological evolution
- Sediment structure
- Physical oceanography

Measurements

- Water currents
- Salinity (conductivity)
- Volcanic vent volatiles
- Visible/sonar imaging
- Temperature, pressure
- Biologicals

Science objectives in the ice

- Presence, effects of biologicals
- Stratigraphy/climate

Measurements

- Profile ice temperature, density, particulates
- Isotopic abundance
- Biologicals

Europan cryobot schematic

Figure 11.6. Cryobot of NASA's Jet Propulsion Laboratory (U.S.A.), which should go from the ice surface of Europa to its "subglacial" sea, melting-sublimating a cavern in the ice body and sinking its way to the underlying water under gravity without maintaining a borehole in the ice behind it. In 1996 NASA also suggested this device for the study of the subglacial Lake Vostok (after Jet Propulsion Laboratory Project, 1996). (1) "Hot-melting drill head" making a hole with a diameter of about 15 cm; (2) heat source modules; producing about 4 kW; (3) hydrobot to study interior of the lake; (4) spool of cable, connected to the surface; and (5) the hole, which will be refrozen behind the cryobot.

This would be possible because the cable connecting the cryobot with the surface would be spooled within the cryobot and would thus be unspooling while the device was sinking, similar to the Philberth probe described in Chapter 3.

The energy requirement for melting ice through the Antarctic Ice Sheet to Lake Vostok is about 5 kW for a vertical speed of about 1.5 m ah hour, with the energy to be transferred from the surface by cable. It was proposed in 1996 that the development and manufacture of a cryobot for Lake Vostok would be made through an award of the Office of Polar Programs of the National Science Foundation (U.S.A.). It was projected that the first tests of this kind of a device would be in deep ice in Antarctica or Greenland, but nearly 8 years passed without much production/manufacuring progress.

Plans for a study of Europa propose that a cryobot should carry all the energy for penetration through the Europan ice cover, with the cable to the surface only being used for communication of information to the surface. In this case the "General Purpose Heat Source" (GPHS) modules of the 16 cm diameter cryobot should carry about as much energy as, for example, is emitted from the burning of about 2,000 kg of kerosene, the amount of energy that can be placed and released in the space shown in Figure 11.6 by nuclear power only. Therefore, it is logical to conclude that nuclear powered reactors would be used for cryobots undertaking planetary exploration (Elliot *et al.*, 2003).

In relation to this, in 1963 Dr. Kapitsa and I proposed a Sub-glacial Autonomous Station (SGAS) for penetration of the Antarctic sheet to study the bottom water layer or a subglacial lake, if one existed. We suggested the use of a small nuclear power plant in the lower part of a hermetically closed container of the SGAS that would include various instruments and equipment. The nuclear power plant had to produce enough energy to melt to the bottom of the ice sheet. No hole was needed because the SGAS was supposed to descend in a water cavern melted by the unit, which would refreeze above it. Communication with the ice sheet surface would be kept by wireless communication. The Atomic Energy Institute of the U.S.S.R. Academy of Sciences approved the project. The Institute agreed to provide a small (diameter about 0.5 m) nuclear energy reactor of 100 kW, small enough to be installed in the SGAS's container, a cylinder with diameter 0.9 m. This power source would provide a calculated speed of vertical penetration through the ice of slightly more than 1 m hr^{-1}. In other words the SGAS could reach the bottom of the Antarctic Ice Sheet below the Vostok Station area in less than 4 months. The device would be installed in a special hole at the surface of the ice sheet and the nuclear reactor would then be turned on.

This project was never realized due to issues with the classified status of the reactor. There was also another obstacle. At this early stage of the project, we did not know how to bring the SGAS back to the surface of the ice sheet – we could not leave it at the bottom because the Antarctic Treaty prohibited the disposal of items like nuclear power plants in Antarctica. After 12 years this idea was discussed again. The problem of bringing the SGAS back to the ice sheet surface was solved in 1975, when I was trying to publish my book on the thermophysics of glaciers. One chapter of the book discussed the possibility of radioactive waste disposal in a central part of the Antarctic Ice Sheet (Weertman *et al.*, 1974; Zeller *et al.*, 1973) from a thermophysical point of view (sinking of containers of radioactive waste to the bottom of the ice sheet due to ice melting because of the nuclear energy dissipation of the waste (Zotikov, 1986)). Everything connected with radioactive waste disposal was classified at this time in the Soviet Union. I went to the very top of the Committee on Management of Nuclear Energy of the Soviet Union and showed my manuscript to the Vice Chairman of this Committee, Dr. Morochov, and we discussed many questions, including the SGAS return problem. In a few minutes he had a solution. The SGAS melts its way down because its weight is greater than the weight of the amount of water it displaces in the meltwater cavern. If, at the bottom of the ice sheet, its weight can be reduced to less than the weight of the water it displaces, then it will start to float in its water cavern. If we then place the heat generated by the nuclear reactor at its topmost surface, the device will start to melt its way up. It would not be a SGAS, but SGARS (Sub-glacial Autonomous Returnable Station) in this case.

This idea had new interest in 1994 (Zotikov, 1993), after the first Cambridge workshop of 1993, when it became clear that a large subglacial lake below Vostok Station existed. The same Atomic Energy Institute (now Kurchatov's Institute) agreed on a suggestion of me to take part in designing a new, different type of unit. The Institute suggested it should be powered by a nuclear thermo-electric plant producing 0.5–2 MW of heat and 100 kW of electricity – capable of doing this for many years in an environment of the type of Lake Vostok (Figure 11.7).

This project for a long-term study of Lake Vostok was presented at the second Cambridge workshop in 1995, but the attitude to the use of nuclear energy for a type of power plant in Antarctica has since changed, and it looks like the use of it in this case would meet opposition. I was told that Vostok Station is located at the sector of Antarctica that Australia claims as its territory. Australia prohibits the use of nuclear energy in its country, and we predict that Australia would be against this project in spite of the fact that all territorial claims in Antarctica are frozen by the Antarctic Treaty.

The idea of the use of nuclear energy for melting through the ice sheet on other planets with the use of a cryobot design was suggested (French *et al.*, 2001) in a project entitled "Palmer Quest: Searching for life below the ice caps of Mars" (private commun., Dr. Cassidy, 2004). Palmer Quest is a major candidate for a NASA mission to the North Polar Cap of Mars in a search for extant life in a basal domain of the ice sheet. Figure 11.8 shows an image of the North Polar Cap of Mars with the general area of the proposed Palmer Quest landing site shown by an oval.

It is believed that the North Polar Cap is formed from a mixture of ice and dust with a thickness of approximately 3 km spreading for hundreds of kilometers away from the Pole. The bottom of this ice cap is much warmer than its surface, just as in the Antarctic, but it was thought for a while that there is no melting and liquid water at the bottom of this ice cap, because of its insufficient thickness (Clifford, 1987). However, there are some features of the ice cap, found recently from a study of the Mars Orbiter Laser Altimeter (MOLA) data, which may be interpreted as an indication of catastrophic water outflow of subglacial water of the North Polar Cap (Fishbaugh and Head, 2002).

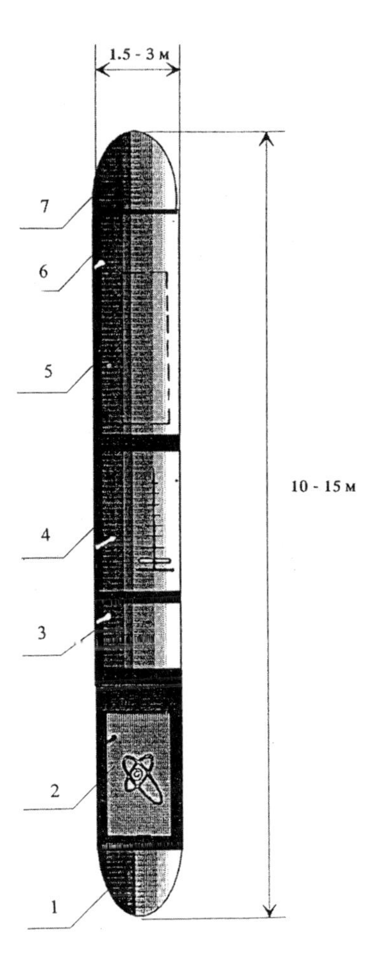

Figure 11.7. Nuclear powered SGARS, which would be capable of going from the surface of the Antarctic Ice Sheet to subglacial Lake Vostok, melting a water cavern in the ice body and sinking its way to the lake, staying there for years and returning to the surface floating in a cavern of meltwater. The author presented this concept at the 2nd Cambridge Workshop on Lake Vostok (after the drawing by Russian Scientific Center "Kurchatov's Institute"). (1) Lower thermal drill; (2) nuclear thermo-electrical unit; (3) commanding block; (4) scientific, navigation and communication equipment block; (5) device for monitoring/storing many years of information about lake bottom processes (water pressure, geothermal heat flow, biological, and other condition changes); (6) ballast compartment – filled with water to make the station sink, devoid of water to make the station surface with the aid of an upper thermal drill (7) for melting the upper part of the cavern.

In a proposal for "Palmer Quest: Searching for life below the ice caps of Mars", Dr. Carsey, Principal Investigator of the proposal writes: "The mission is named for Nathaniel B. Palmer, a bold 21 year old American sea captain who sailed his 47 ft (14.2 m) sailboat, *Hero*, in search of seals, to the coast of Antarctica in November

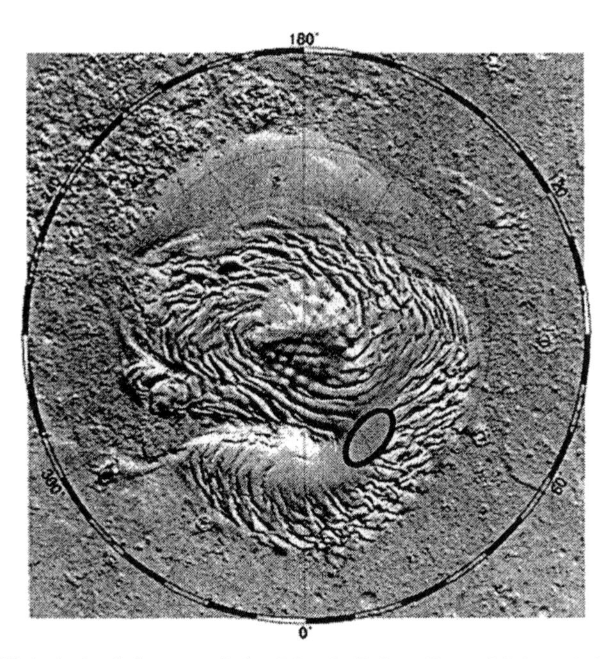

Figure 11.8. MOLA-derived image of the North Polar Cap of Mars (after Carsey, 2004, private commun.). The oval in the picture shows an area where the search for life in the ice is planned as a part of the Palmer Quest mission.

1820, becoming the first person to sight the landmass of Antarctica at what is now Palmer Land. We plan to follow his example to fearfully search for life on Mars; we do not know what we will discover."

Epilog

On a windy, cold day, 11 November 2005, a large Russian ice-breaking ship *Academician Federov* left the port of St. Petersburg and began the month-long trip to Antarctica. Giving his best wishes to those on board, Dr. V. Lukin, Head of the Russian Antarctic Expedition, made it clear that one of the most important goals of this particular expedition would be the beginning of a new step in the long saga that is the deep core drilling at Vostok Station. The expedition would oversee the drilling of the first 50 m of the last 120 m of ice at the bottom of the Vostok Station borehole, with the intention of reaching the boundary of Lake Vostok and penetrating the Lake.

For me, an old actor in this play, these words meant that people who had studied Lake Vostok for a long time and had spent more that 30 years core drilling more than 90% of the 3,500 m of ice cover, separating the Lake from our present-day environment, and had extracted thousands of meters of biologically clean ice cores, having waited for 7 years after this non-stop core drilling for fear of "contaminating the Lake", need wait no longer. A project had been designed with new devices, capable of completing the final drilling with as little contamination as possible. The idea behind the project had been open to the scientific community for years of evaluation, resulting in many positive responses. However, an article in *Science* (Inman, 2005), published just a month before *Academician Federov* left port, was not supportive of the idea. The Russian Antarctic Expedition supported by its own, and other, experts' positive remarks decided not to wait any longer and began drilling, aiming to complete penetration within a few years.

At the end of February 2006 drilling of approximately the first 27 m of the borehole was complete, with the depth of the borehole being increased from 3,623 m to 3,650 m. Each round of drilling brought to the surface about 1.6 m of core and took approximately 6 hours of drilling time.

Everything has gone smoothly and the core of ice taken to the surface. Part of it has been sent back to Russia for study, while another part is stored at Vostok Station (below −56°C).

Presently, less than 100 m of ice remains above the Lake. I believe that the first important chapter of the scientific thriller "Antarctic Subglacial Lake Vostok", will soon be complete. In about 2 years, at the end of the International Polar Year (2007/2008 Antarctic season), man-made equipment will make contact with Lake Vostok's water.

I believe that a result of this will be a rapid increase in interest in everything connected with the Lake, which arguably represents the greatest geographical discovery of the second part of the last century. I hope this book helps them to understand the historical and present-day situation at Lake Vostok.

Bibliography

Abizov, S. S., 1993. Micro-organisms in the Antarctic ice. In: *Antarctic Microbiology* (Friedmann, E. I., ed.). Wiley–Liss, New York, pp. 265–295.

Abizov, S. S., 2001. Microflora of the Central Antarctica Ice Sheet. DSc. thesis, Institute of Microbiology, Russian Academy of Sciences, Moscow, 64pp [in Russian].

Abizov, S. S., Mitskevich, I. N., Bobin, N. E., Koudryshov, B. B., and Pashkevich, V. M., 1994. Longevity of micro-organisms in the most ancient layers of the central Antarctic ice sheet. *Proceedings, 30th COSPAR Symposium, Hamburg, 16 July, 1994*, p. 319.

Abizov, S. S., Mitskevich, I. N., Poglazova, N. I., Barkov, N. I., Lipenkov, V. Ya., Bobin, N. E., Koudryshov, B. B., and Pashkevich, V. M., 1998a. Microflora in the ice sheet strata of central Antarctica above the subglacial lake in the vicinity of Vostok Station. *Lake Vostok Study: Scientific Objectives and Technological Requirements, March 24–26, 1998*. Arctic and Antarctic Research Institute, St. Petersburg, Russia (Abstracts, pp. 45–46).

Abizov, S. S., Mitskevich, I. N., and Poglazova, M. N., 1998b. Microflora of the deep glacier horizons of Central Antarctica. *Microbiology*, **67**(4), 451–458 [in Russian].

Abizov S. S., Mitskevich, I. N., and Poglazova, N. I., 2001. Microflora in the basal strata of Antarctic ice core above Lake Vostok. *Advances in Space Research*, **28**, 701–706.

Alekhina, I. A., Petit, J. R., Lukin, V. V., Vasiliev, N. I., and Bulat, S. A., 2005. Estimate for bacterial contents of 5G-1 borehole drilling fluid, Vostok Station, Antarctica. *Data of Glaciological Studies*, **98**, 109–117.

Atlas of Antarctica, 1966. (Volume 1). Main Department of Geodesy and Mapping, Leningrad, 310 pp [in Russian].

Bamber, J. L., 1994. A digital elevation model of the Antarctic ice sheet derived from ERS-1 altimeter data and comparison with terrestrial measurements. *Annals of Glaciology*, **20**, 48–54.

Barkov, N. I., Vostretsov, R. N., Lipenkov V. Ya., and Salamatin, A. N., 2002. Air temperature and precipitation variations in Vostok Station area through four climatic cycles during recent 420,000 years. *Arctic and Antarctic*, **1**(35), 82–97 [in Russian].

Bell, R. E. and Karl, D. M., 1998. *Lake Vostok Workshop – Lake Vostok: A Curiosity or a Focus for Interdisciplinary Study*? Washington D. C., pp. 2–11.

Bell, R. E., Studinger, M., Tikku, A. A., Clarke, G. K. C., Gutner, M. M., and Meertens, C., 2002. Origin and fate of Lake Vostok water frozen to the base of the East Antarctic ice sheet. *Nature*, **416**, 307–310.

Bentley, C., 1996. Antarctica: Water kept liquid by warmth from within. *Nature*, **381**(6584), 644–646.

Blankenship, D. D., Bentley, C. R., Rooney, S. T., and Alley, R. B., 1986. Seismic measurements reveal a saturated porous layer beneath an active Antarctic ice stream. *Nature*, **322**, 54–57.

Browning, J. A., 1978. Flame-drilling through the Ross Ice Shelf. *The Northern Engineer*, **10**(1) Spring, 10–16.

Browning, J. A., Bigl, R. A., and Sommerville, D. A., 1979. Hot water drilling and coring at site J-9, Ross Ice Shelf. *Antarctic Journal of the United States*, **XIV**(5), October, 73–76.

Bulat, S. A., Alekhina, I. A., and Lukin, V. V., 2001. Looking for microbes in Lake Vostok, Antarctica: The case of basal (Accreted) ice cores. *VIII SCAR International Biology Symposium "Antarctic Biology in a Global Context", 27 August–1 September 2001*. Vrije Universiteit, Amsterdam, The Netherlands (Abstract S3O23).

Bulat, S. A., Alekhina, I. A., Blot, M., *et al.*, 2004. DNA signature of thermophilic bacteria from the aged accretion ice of Lake Vostok. Antarctica: Implications for searching life in extreme icy environments. *International Journal of Astrobiology*, **3**, 1–12.

Cano, R. and Borucki, M., 1995. Revival and identification of bacterial-spores in 25-million-year-old to 40-million-year-old Dominican amber. *Science*, **268**(5213), May, 1060–1064.

Carsey, F. D. and Horvath, J. C., 1996. A program concept for integrated studies of ice and oceans on Earth and Europa. *Earth/Europa Workshop White Paper, October*.

Carsey, F. D., Cutts, J., and Horvath, J. C., 1998. The NASA–JPL Europa–Lake Vostok initiative. *Lake Vostok Study: Scientific Objectives and Technological Requirements, March 24–26, 1998*. Arctic and Antarctic Research Institute, St. Petersburg, Russia (Abstracts, pp. 70–71).

Carsey, F. J., Chen, J., Cutts, L., French, R., Kern, A. L., Lane, A., Stolorz, W., Zimmerman, V., and Ballou, P., 2001. The profound technical challenges inherent in exploring Europa's hypothesized ocean. *Marine Technology Society Journal*, **33**, 28–32.

Clifford, S., 1987. Polar basal melting on Mars. *J. Geophys. Res.*, **92**(B9), 9135–9152.

Clow, G. D. and Koci, B., 2002. A fast mechanical-access drill for polar glaciology, paleoclimatology, geology, tectonics, and biology. *Memoirs of National Institute of Polar Research Special Issue No. 56*. Ice Drilling Technology, Tokyo.

Dansgaard, W., 1964. Stable isotopes in precipitation. *Tellus*, **16**, 436–468.

Dansgaard, W., Johnsen, S. J., Clausen, H. B., and Gundestrup, N., 1973. Stable isotope glaciology. *Meddelelser Gronland*, **197**(2), 53.

Doran, P. T., 1998. Perennially ice-covered and deeply ice-covered lakes in the McMurdo dry valleys: lessons for Vostok. *Lake Vostok Study: Scientific Objectives and Technological Requirements, March 24–26, 1998*. Arctic and Antarctic Research Institute, St. Petersburg, Russia (Abstracts, pp. 56–58).

Ellis-Evans. J. C., 1998. Aquatic microbial microsystems of Antarctica: Biodiversity in adversity (pointers for life in Lake Vostok?). *Lake Vostok Study: Scientific Objectives and Technological Requirements, March 24–26, 1998*. Arctic and Antarctic Research Institute, St. Petersburg, Russia (Abstracts, pp. 50–55).

Ellis-Evans, J. and Wynn-Williams, D., 1996. Antarctica: A great lake under the ice. *Nature*, **381**(6584), 644–646.

Elliot, J. O., Lipinski, R. J., and Poston, D. I., 2003. Mission concept for a nuclear reactor-powered Mars cryobot lander. *Proceedings of Space Technology and Applications Inter-*

national Forum (Space Technology and Application International Forum 2003), AIP Conference Proceedings, p. 654.

Fishbaugh, K. E., and Head, J. W., 2002. Chasma Boreale. Mars: Topographic characterisation from MOLA data and implications for mechanisms of formation. *J. of Geophys. Res.*, **107**(E3), 10, 1029/2000Je001351.

French, L. F., Anderson, F., Carsey, F., French, G., Lane, A., Shakkottay, J., Zimmerman, V., and Engelhardt, H., 2001. Cryobots: An answer to subsurface mobility in planetary icy environments. *Proceedings of the International Symposium on Artificial Intelligence, Robotics and Automation in Space, Montreal*, p. 30.

Fowler, S. H., Feechan, M., and Berning, S., 1999. Development update and applications of an advanced-composite spoolable tubing. *Offshore Technology Conference, Houston, TX, 4–7 May* (Paper 8621).

Gaddy, D. E., 2000. Coiled-tubing drilling technologies target niche markets. *Oil & Gas Journal*, **98**(2), 10.

Giles, J., 2004. Russian bid to drill Antarctic lake gets chilly response. *Nature*, **430**(6999), July, 494–495.

Goldstein, R. M., Engelhardt, H., Kamb, B., and Frolich, R. M., 1993. Satelite radar interferometry for monitoring ice-sheet motion-application to an Antarctic ice stream. *Science*, **262**(5139), December, 1525–1530.

Gilichinsky, D. A., Wiagner S., and Vishnevskaya, A., 1995. Permafrost microbiology. *Permafrost and Periglacial Processes*, **6**, 281–291.

Glebovsky, Yu. S., Karasik, A. M., and Lastochkin, V. M., 1966. Magnetic anomalies from airborne investigations. Part I: East Antarctica, 1:7,000000. *Atlas of Antarctica* (Volume 1). Moscow-Leningrad, pp. 40–41.

Greenberg, R. (2005) *Europa – An Ocean Moon: Search for an Alien Biosphere*. Springer–Praxis, Chichester, U.K.

Grushinsky, N. P., Koryakin, E. D., Stroev, P. A., Lazarev, G. E., Sidorov, D. V., and Virskaya, N. F., 1972. Catalogue of gravity data points of Antarctica. *Proceedings of P.K. Sternberg State Astronomic Institute, Moscow State University*, **XLII**, 115–311 [in Russian].

Ignatov, V. S., 1960. The thermal drilling of boreholes at Vostok Station. *Inform. Bull. of Soviet Antarctic Expedition*, **22**, 31–34.

Inman, M., 2005. The plan to unlock Lake Vostok. *Science*, **310**, October, 611–612.

Izvestiya, 1963. Interview on reader's request: Does it melt or does it not melt? *Izvestija*, **112**(1399), p. 3 [in Russian].

Jouzel, J., Barkov, N. I., Barnola, J. M., Bender, M., Chapellaz, J., Genton, C., Kotlyakov, V. M., Lipenkov, V. Ya., Lorius, C., Petit, *et al.*, 1993. Extending the Vostok Ice core record of paleoclimate to the penultimate glacier period. *Nature*, **364**(6436), July, 407–412.

Jouzel, J., Petit, J. R., Souchez, R., Barkov, N. I., Lipenkov, V. Ya., Raynaud, D., Stievenard, M., Vasiliev, N. I., Verbeke, V., and Vimeux, F., 1999. More than 200 meters of lake ice above subglacial lake Vostok, Antarctica. *Science*, **286**, 10 December, 2138–2141.

Kapitsa, A. P., 1961. Dynamics and morphology of a glacier ice sheet of central sector of East Antarctica. *Proceedings of Soviet Antarctic Expedition*, **18**, 93 pp [in Russian].

Kapitsa, A. P., 1968. The subglacial topography of Antarctica. Nauka Publishing, Moscow, 162 pp [in Russian].

Kapitsa, A. P., Ridley, J. K., Robin, G. de Q., Siegert, M. J., and Zotikov, I. A., 1996. A large deep freshwater lake beneath the ice of central East Antarctica. *Nature*, **381**(6584), 684–686.

Karl, D. M., Bird, D. F., Bjorkman, K., Houlihan, T., Sgakelford, R., and Tupas, L., 1999. Micro-organisms in the accreted ice of Lake Vostok, Antarctica. *Science*, **286**, 2144–2147.

Kennicut, M., 2001. *Report on the Subglacial Antarctic Lake Exploration Group of Specialists (SALEGOS)*. Meeting 1 in Bologna, Italy, November 2001, 69 pp.

Kennicut, M., 2002a. *Report on the Subglacial Antarctic Lake Exploration Group of Specialists (SALEGOS)*. Meeting 2 at Lamont-Doherty Earth Observatory, USA, October 2002, 21 pp.

Kennicut, M., 2002b. *Report on the Subglacial Antarctic Lake Exploration Group of Specialists (SALEGOS)*. Meeting 3 at University of California, Santa Cruz, USA, May 2002, 21 pp.

Kennicut, M., 2003a. *Report on the Subglacial Antarctic Lake Exploration Group of Specialists (SALEGOS)*. Meeting 4 in Chamonix, France, April 2003, 17 pp.

Kennicut, M., 2003b. *Report on the Subglacial Antarctic Lake Exploration Group of Specialists (SALEGOS)*. Meeting 5 in Bristol, UK, October 2002, 15 pp.

Kennicut, M., 2002b.

Koci, B., 1999. The AMANDA Project: Drilling precise, large diameter holes using hot water. *Memoirs of National Institute of Polar Research Special Issue No. 49*. Ice Drilling Technology, Tokyo, pp. 203–211.

Kropotkin, P. A., 1876. *Investigation of Quaternary Period*. Notes of Russian Geographical Society (no. 6)., St. Petersburg, 717 pp [in Russian].

Kotlyakov, V. M. and Lorius, C., 2000. Four climatic cycles based on ice core data from deep drilling at the Vostok Station, Antarctica. *Polar Geography*, **24**(1), 35–52.

Kudryashov, B. B., Chistyakov, V. K., Zagrivny, E. A., and Lipenkov, V. Ya., 1982. Preliminary results of deep drilling at Vostok Station, Antarctica 1981–1982. *Proceedings of the Second International Workshop/Symposium on Ice Drilling Technology, Calgary, Alberta, Canada 30–31 August 1982*. U.S.A. Cold Regions Research Engineering Laboratory Special Report, pp. 123–124.

Kudryashov, B. B., Vasiliev, N. I., Vostretsov, R. N., Zubkov, V. M., Krasilev, A. V., Talalay, P. G., Barkov, N. I., Lipenkov, V. Ya., and Petit, J. A., 2000. Deep ice coring at Vostok Station (East Antarctica) by electromechanical drill. *International Workshop on Ice Drilling Technology, Nagaoka University of Technology, Japan, October 30–November 1, 2000*, pp. 21–31.

Lipenkov, V. Ya. and Barkov, N. I., 1998. Internal structure of the Antarctic ice sheet as revealed by deep core drilling at Vostok Station. *Lake Vostok Study: Scientific Objectives and Technological Requirements, March 24–26, 1998*. Arctic and Antarctic Research Institute, St. Petersburg, Russia (Abstracts, pp. 31–35).

Lipenkov, V. Ya. and Istomin, V. A., 2001. On the stability of air clathrate–hydrate crystals in subglacial Lake Vostok, Antarctica. *Material of Glaciological Investigations*, **91**, 138–148.

Leitchenkov, G. L., Verkulich, S. R., and Mosolov, V. N., 1998. Tectonic setting of Lake Vostok and possible information contained in its bottom sediments. *Lake Vostok Study: Scientific Objectives and Technological Requirements, March 24–26, 1998*. Arctic and Antarctic Research Institute, St. Petersburg, Russia (Abstracts, 62–65).

Masolov, V. N., Lukin V. V., Sheremetiev, A. N., and Popov, S. V., 2001. Geophysical investigations of the subglacial Lake Vostok in East Antarctica. *Proceedings of Russian Academy of Sciences*, **379A**, 680–685.

Masolov, V. N., Kudryavzev, G. A., Sheremetiev, A. N., Popkov, S. V., Lukin, V. V., Grikurov, G. E., and Leitchenkov, G. L., 1999. Earth science studies in the Lake Vostok region: Existing data and proposals for future research. *Scientific Committee on Antarctic Research International Workshop on Subglacial Lake Exploration. Cambridge, England, September 1999*. Workshop Report and Recommendations, pp. 1–18.

Mayer, C., Grosfeld, K., and Siegert, M. J., 2003. The effect of salinity on water circulation within subglacial Lake Vostok. *Geophysical Research Letters*, **30**(14).

McIntyre, N. F., 1983. The topography and flow of the Antarctic Ice Sheet. A dissertation, submitted for the degree of Dr. of Philosophy in the University of Cambridge, St. John College, Cambridge, November, 170 pp.

Montagnat, M., Duval, P., Bastie, P., Hamelin, B., Brissand, O., de Angelis, M., Petit, J. R., and Lipenkov, V. Ya., 2001. High crystalline quality of large single crystals of subglacial ice above Lake Vostok (Antarctica) revealed by hard X-ray diffraction. *Série II Fascicute a – Sciences de la Terre et des Planètes*, **333**, 419–425.

Muir, H., 1996. Giant lake lurks beneath Antarctic ice. *New Scientist*, 20 June.

Nuttall, N., 1996. Vast lake discovered beneath the ice of Antarctica. *The Times*, Thursday 20 June.

Nye, J. F., 1959. Motion of ice sheets and glaciers. *Journal of Glaciology*, **3**, 493–507.

Oswald, G. K. L. and Robin, G. de Q., 1973. Lakes beneath the Antarctic Ice Sheet. *Nature*, **245**, 251–254.

Petit, J. R., Basile, I., Jouzel, J., Barkov, N. I., Lipenkov, V. Ya., Vostretsov, R. N., Vasiliev, N. I., and Rado, C., 1998. Preliminary investigations and implications from the 3,623 m Vostok deep core studies. *Lake Vostok Study: Scientific Objectives and Technological Requirements, March 24–26, 1998*. Arctic and Antarctic Research Institute, St. Petersburg, Russia (Abstracts, pp. 43–45).

Petit, J. R., Basile, I., Leruyuet, A., Raynaud, D., Lorius, C., Jouzel, C., Stievenard, M., Lipenkov, V. Ya., Barkov, N. I., Kudryashov, *et al.*, 1997. Four climate cycles in Vostok ice core. *Nature*, **387**(6631), 359–360.

Philberth, K., 1974. The Thermal Probe Deep Drilling Method by EGIG in 1968 at Station Jarl-Joset, central Greenland. In: J. F. Splettstoesser (ed.), *Proceedings of the Ice Core Drilling Symposium*. University of Nebraska Press, pp. 117–121.

Popkov, A. M., Kudryavsev, G. A., Verkulich, S. N., Masolov, S. N., and Lukin, V. V., 1998. Seismic studies in the vicinity of Vostok Station (Antarctica). *Lake Vostok Study: Scientific Objectives and Technological Requirements, March 24–26, 1998*. Arctic and Antarctic Research Institute, St. Petersburg, Russia (Abstracts, pp. 26–27).

Popov, S. V., Mironov, A. V., and Sheremetiev, A. N., 2000. Results of ground-based radio-echo survey of the subglacial Lake Vostok in 1998–2000. *Materials of Glaciological Iinvestigations*, **89**, 129–133 [in Russian].

Priscu, J. C., Adams, E. E., Lyons, W. B., Voytek, M. A., Mogk, D. W., Brown, R. L., McKay, C. D., Tokacs, C. D., Welch, K. A., and Wolf, C. E., 1999. Geomicrobiology of subglacial ice above Lake Vostok, Antarctica. *Science*, **286**, 2141–2144.

Priscu, J. C., Bell, R. E., Bulat, S. A., Ellis-Evans, J. C., Kennicut II, M. C., Lukin, V. V., Petit, J.-R., Powell, R. D., Sieger, M. J., and Tabacco, I. A., 2003. International plan for Antarctic subglacial lake exploration. *Polar Geography*, **27**(1), 289–301.

Psener, R. and Sattler, B., 1999. Food webs in Lake Vostok? Hypotheses about a hidden ecosystem. *Scientific Committee on Antarctic Research International Workshop on Subglacial Lake Exploration, Cambridge, England September 1999*. Workshop Report and Recommendations, pp. 43–49.

Quigg, P. W., 1983. *A Pole Apart. The Twentieth Century Fund*. McGraw Hill, 299 pp.

Ratford, T., 1996. Polar lake may hold "lost world". *The Guardian*, Thursday 20 June.

Ridley, G. P., Gudlip, W., and Laxon, S. W. 1993. Identification of subglacial lakes using ERS-1 radar altimeter. *Journal of Glaciology*, **339**(133), 623–634.

Robin, G. de Q., 1955. Ice movement and temperature distribution in glaciers and ice sheets. *Journal of Glaciology*, **2**(18), 523–532.

Robin, G. de Q., Drewry, D. J., and Meldrum, D. T., 1977. International studies of the ice sheets and bedrock. *Phil. Trans. R. Soc. Lond. B.*, **279**, 185–196.

Robinson, R. V., 1960. From a visual navigation experience in flights in Antarctica. *Inform. Bull. Soviet Antarctic Exped.*, **18**, 28–29 [in Russian].

Roura, R., 1999. Subglacial lake exploration and the Lake Vostok case: To drill or not to drill? *Scientific Commitee on Antarctic Research International Workshop on Subglacial Lake Exploration, Cambridge, England September 1999*. Workshop Report and Recommendations, pp. 21–24.

Salamatin, A. N., 1998. Modeling ice sheet dynamics and heat transfer in central Antarctica at Vostok station. Conditions of subglacial lake existence. *Lake Vostok Study: Scientific Objectives and Technological Requirementsm, March 24–26, 1998*. Arctic and Antarctic Research Institute, St. Petersburg, Russia (Abstracts, pp. 36–37).

Salamatin, A. N., 2000. Paleoclimatic reconstructions based on borehole temperature measurements in ice sheets: Possibilities and limitations. In: T. Hondoh (ed.), *Physics of Ice Core Records*. Hokkaido University Press, pp. 243–282.

Siegert, M. J., 2001. Subglacial lake and deep ice exploration: Canadian expertise and international opportunities. In: O. H. Loken and N. Couture (eds), *The Physiography of modern Antarctic Subglacial Lakes* (Report on the International workshop). Canadian Polar Commission Ottawa, pp. 8–10.

Siegert, M. J., Dowdeswell, J. A., Gorman, M. R., and McIntryre, N. F., 1996. An inventory of Antarctic sub-ice lakes. *Antarctic Science*, **8**, 281–286.

Siegert, M. J., Tranter, M., Ellis-Evans, J. C., Priscu, J. C., and Berry, W., 2003. The hydrochemistry of Lake Vostok and the potential for life in Antarctic subglacial lakes. *Hydrological Processes*, **17**, 795–814.

Souchez, R., Petit, J. R., Tison, J.-L., Jouzel, J., and Verbeke, V., 2000. Ice formation in subglacial Lake Vostok, central Antarctica. *Earth and Planetary Science Letters*, **181**, 529–538.

Ueda, H. T. and Garfield, D. E., 1969. *Core Drilling Through the Antarctic Ice Sheet* (Technical Report 231, December 1969). U.S. Army Cold Regions Research Engineering Laboratory, Hanover, NH, 17 pp.

Ueda, H. T. and Garfield, D. E., 1968. *Core Drilling Through the Antarctic Ice Sheet* (U.S.A. Cold Regions Research Engineering Laboratory Special Report 231). U.S. Army Cold Regions Research Engineering Laboratory, Hanover, NH, 25 pp.

Ueda, H. T. and Hansen, B. L., 1967. Installation of deep core drilling equipment at Byrd Station (1966–1967). *Antarctic Journal of the U.S.*, **2**(4), 101–102.

Vorobieva, E. A., Gilichinskiy, D. A., Soina, V. S., Matikelashvili, A., Vishnevetskaia, T. A., Kozhevich, P. A., Polyanskaia, L. M., and Zvygintsev, D. G., 1998. Antarctic permafrost as microbial habitat. *Lake Vostok Study: Scientific Objectives and Technological*

Requirements, March 24–26, 1998. Arctic and Antarctic Research Institute, St. Petersburg, Russia (Abstracts, pp. 47–49).

Verculich, *et al.*, 2002. Proposal for penetration end exploration of sub-glacial Lake Vostok, Antarctica. In: *Memoirs of National Institute of Polar Research* (Special Issue No. 56). Ice Drilling Technology 2000, p. 245–252.

Weertman, J., Siebert, J., Weeks, W. F., and Sternig, J., 1974. Radioactive wastes on ice. *Science Public Affairs. Bulletin of Atomic Scientists*. **30**(1), 30–35.

Williams, M. J. M., 2001. Application of three dimensional model to Lake Vostok: An Antarctic subglacial lake. *Geophysical Research Letters*, **28**, 531–534.

Wilson, E., 1982. The southern journey, Summer 1902–1903. In: H. King (ed.), *South Pole Odyssey*. Blandiford Press, Poole, Dorset, pp. 25–80.

Wuest, A. and Carmack, E., 2000. A priory estimates of mixing and circulation in the hard-to-reach water body of Lake Vostok. *Ocean Modelling*, **2**, 29–43.

Zeller, E., Saunders, D. F., and Angino, E. E., 1973. Putting radioactive wastes on ice: A proposal for an international radionuclide depository in Antarctica. *Science Public Affairs. Bulletin of Atomic Scientists*, **29**(1), 4–9 and 50–52.

Zotikov, I. A., 1959. Experimental study of melting by a hot supersonic flow. *Meteoritica*, **XVII**, 85–91 [in Russian].

Zotikov, I. A., 1961. Heat regime of central Antarctica ice sheet. *Information Bulletin of Soviet Antarctic Expedition*, **28**, 16–21 [in Russian].

Zotikov, I. A., 1962. Heat regime of central Antarctica Ice Sheet. *Antarctica. Commission Report Antarctica*. Doklady Komissii 1961, Vip. 23, Moscow 27–40 [in Russian].

Zotikov, I. A., 1963. Bottom melting in the central zone of Antarctic's ice sheet and its influence on modern balance. *International Association of Scientific Hydrology Publication, Colloque d'Obergurgle, Belgique*, **1**, 36–44.

Zotikov, I. A., 1968. Heat regime of an Antarctic Ice Sheet. DSc. thesis. Arctic and Antarctic Research Institute, Leningrad, 350 pp [in Russian].

Zotikov, I. A., 1977. *Thermal Regime of the Antarctic Ice Sheet*. Leningrad, Gidrometeoizdat, 168 pp [in Russian].

Zotikov, I. A., 1979. *Antifreeze thermodrill for core through the central part of the Ross Ice Shelf (J-9 camp), Antarctica*. USA Cold Regions Research Engineering Laboratory Report, 79–24, 21 pp.

Zotikov, I. A., 1986. *Themophysics of Glaciers*. D. Reidel Publishing Company, a member of Kluwer Academic Publishers Group, Dordrecht/Boston/Lancaster/Tokyo, 275 pp.

Zotikov, I. A., 1993. Subglacier Autonomous Station Project. *The Fourth International Workshop on Ice Drilling Technology, April 20–23, Abstracts*. KKR Takebashi, Tokyo, 46 pp.

Zotikov, I. A., 1998. Lake Vostok, Antarctica (glaciological, biological, planetary aspects). Materials of Glaciological Investigations. Chronicle, Discussions. Vip. 85, 137–147 [in Russian].

Zotikov, I. A. and Duxbury, N. S., 2000. On genesis of Lake Vostok. *Proceedings of Russian Academy of Sciences*, **372**(6), 1–6.

Zubov, N. N., 1955. Some peculiarities of thick ice covers. *News of Moscow State University*, **2**(3), 3–14 [in Russian].

Zubov, N. N., 1956. *About the ice of Arctic and Antarctic*. Moscow State University Publishing House, 118 pp [in Russian].

Zubov, N. N., 1959. Limits of thickness of sea ice and grounded ice. *Meteorology and Hydrology*, **2**, 22–27 [in Russian].

Index

Printing: Mercedes-Druck, Berlin
Binding: Stein + Lehmann, Berlin